CONSTRUCTION DOCUMENTS USING SKETCHUP PRO 2020

WHO SHOULD USE THIS BOOK?

This book is for anyone involved in creating documentation for Construction whether **architect, builder, engineer, interior designer or DIY enthusiast**.

It is recommended that the user should become familiar with SketchUp's tools and concepts before attempting the exercises in this book. While the lessons are quite self-contained, **familiarity with modelling, navigating the model space, and model management is strongly advised. For excellent advice and links to further resources to help you regarding this, please visit the Resources Section on** www.SketchUp.Expert

WHAT DO I NEED TO COMPLETE THIS COURSE?

To make use of this book, you'll need access to **SketchUp Pro 2019** or later. A modern laptop or desktop computer running Windows 7+ or later MacOS X.12 or newer is required to run the software. There are also hardware requirements that you should be aware of.

Hardware and software requirements are available on the SketchUp website.

 https://help.sketchup.com/en/sketchup/sketchup-hardware-and-software-requirements

You'll need an internet connection to download the files that go with this book- (See Link on Page 4.)

CONSTRUCTION DOCUMENTS USING SKETCHUP PRO 2020

CONTENTS

Contents	Page No.	Subsection	Page No.
Who should use this book?			
Contents			
Why SketchUp?			
SketchUp & BIM			
Introduction	1		
Getting Started	2		
Drawing Construction Lines	3		
Drawing The Plan	4 – 18		
Creating Groups & Layers	19 – 21		
Modelling the House	22 – 54		
		Shortcut No. 1: Hide	25
		Structuring Your Model	29 - 30
		Fixing Geometry	33
		Shortcut No. 2: Colour by Axis	44
The Section Toolbar	55		
Setting Up Your Drawings	56 – 78		
		Shortcut No. 3: Previous Views	80
House Model- FinishUp 1	79 - 90		
		Shortcut No. 4: Paste in Place	83
		Shortcut No. 5: Toggle Perspective	84
Modelling Details	91 – 97		
Drawing Production - LayOut	98 – 101		
LayOut: Dimensions	102 – 103		
LayOut: Shapes & Infill	104		
LayOut: Text	105 – 106		
LayOut: Sheet Organisation	107		
LayOut: Model Settings	108		
LayOut: Scaled Drawing	109		
LayOut: Pages & Layers	110		
LayOut: Finishing Sheets 1	111		
LayOut: Scrapbook	112		
LayOut: Finishing Sheets 2	113		
LayOut: Details	114		
Modelling – FinishUp	115		
Acknowledgments & Thanks	116		

CONSTRUCTION DOCUMENTS USING SKETCHUP PRO 2020

WHY IS SKETCHUP BETTER THAN 2D CAD?

SketchUp delivers an efficient 3D model that provides updatable drawings directly from the model. 2D CAD

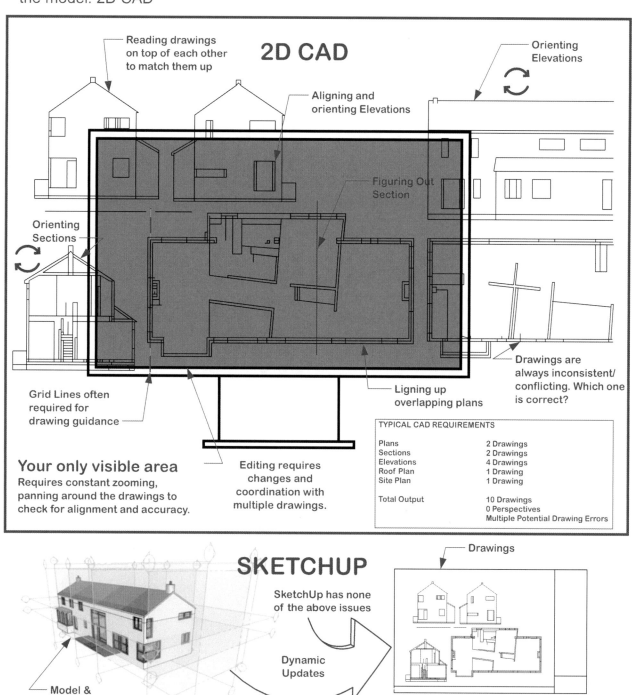

Copyright Paul J. Lee 2020

Why SketchUp?

CONSTRUCTION DOCUMENTS USING SKETCHUP PRO 2020

SKETCHUP AND BIM

SketchUp's unique blend of user friendliness, flexibility and extendability means that it has quite an unlimited potential as a BIM tool.
Facility modelled in SketchUp using BIM functionality (Attribution: sketchupitalia.it)

While certain tasks are handled better by other platforms, they are limited to a particular speciality. Not so with SketchUp, which can be used for a myriad of purposes. For example:

- 3D virtual tours
- Construction sequencing
- Cost reports and analysis
- Energy analysis/ strategy appraisal
- Photographic Renders (with extensions)
- Clash Detection
- Online Collaboration
- Component Cataloguing
- Parametric Modelling
- IFC Classification
- Construction Documentation (CAD)
- Open Program Interface (Extendability)
- Shadow studies

This list is by no means exhaustive.

SketchUp is not hampered by large file sizes or complex automated ("Parametric") components. Modelling an object doesn't require programming an object or looking at a dialogue box. Sharing files is instantaneous. SketchUp Pro is used as a BIM tool by both small and large organisations ranging from one-man operations to multinationals either as a complete business solution or part of a set of tools.

See SketchUp.Expert Resources for Examples of SketchUp for BIM in action:

- **Mortenson Construction**: SketchUp Pro in construction - video case study
- **Barton Malow** 3D for Construction: delivering quality & efficiency
- **McCarthy Building Company**: Case Study

Find out more at the **BIM** section on www.SketchUp.Expert

CONSTRUCTION DOCUMENTS USING SKETCHUP PRO 2020

INTRODUCTION

This is the second version of the book entitled "Construction Documents Using SketchUp Pro & LayOut." which was released in October 2012 as a digital download from **Sketchucation.com**. It was the first book ever released on the topic of producing Construction Documents from SketchUp Pro. Since that time a number of great titles have been published by other authors around the same topic. Following the digital rollout, this book was published in paper format in 2013 on Createspace (Now Kindle Direct.) A lot has happened with SketchUp since the book's release. There have been some excellent improvements in the software, tools and plugins. This update is well overdue.

This course is a practical step by step course showing you how a building model is created and organised. **First I'll show you what we're going to build, then we'll go back to start to building it from scratch.**

Visual connection between front and garden — Garden Perspective (East)

The project I've selected for these training exercises is a small home refit/ extension that I designed and project managed in 2011. The location is a suburb of Cork City, Ireland. I selected this project because it's small in size but reasonably complex- presenting us with a worthwhile learning opportunity.

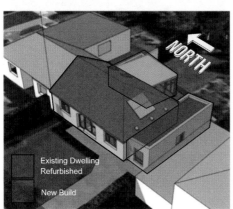

This view from the South West shows the existing vs new elements.

TECHNICAL DESCRIPTION

All interior walls were demolished except for the central structural wall which supports the roof. Existing floors and walls were upgraded and timberframe extensions were added to the East and South. The new plan exploits the split level of the site, creating a generously-sized sun room to the East facing garden.

I've been teaching SketchUp since 2008 and having tried out different approaches I've found the "dive-in-first" to be the best- Building starts immediately and then after a while we review what we've been doing- We're ready for a little bit of theory. After looking at some tools and settings we continue building some more and then review again etc. At the end we finish with tangible assets: A set of 3D models and drawings that look great in your portfolio.

I hope you'll enjoy this course and hope that SketchUp brings benefits to you as it's done for me.

Paul Lee.

CONSTRUCTION DOCUMENTS USING SKETCHUP PRO 2020

GETTING STARTED

-> RESOURCES PAGE

First let's check that we have the companion course models for this manual:

   A. Construction+Documents+Using+SU+Pro+2020_House+Plan+X.X.skp
   B. Details_X.X+Construction+Documents+SketchUp+Pro+2020_(For+All+Users).skp

If you don't have the above SketchUp files, you'll need to download them from the **Resources Section at www.SketchUp.Expert**

To begin, open the SketchUp file "Construction Documents SketchUp Pro House Plan_X.X". You'll see the "Scene" as illustrated in (1). This provides instructions regarding the use of the file. Clicking on the "Plan" Scene as directed (2), you'll notice that it consists of an outline drawing and an image. First we're going to finish the highlighted part of the plan shown below. (The rest is pre-drawn.)

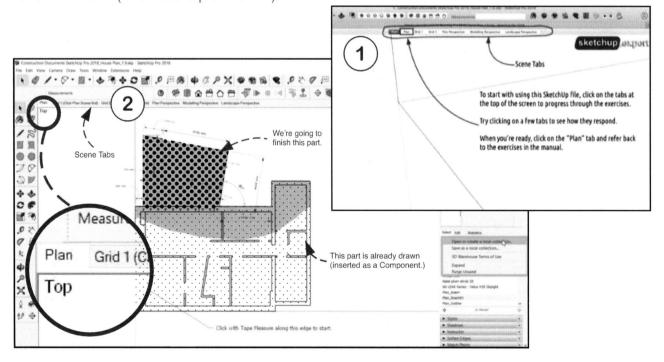

Check out the Scene Tabs to see what each of them contains. When you've finished exploring, click on the "Plan" tab to begin the exercises. The units used in this book are **millimetres**- Our SketchUp files have a millimetre default setting.

CONSTRUCTION DOCUMENTS USING SKETCHUP PRO 2020

DRAWING CONSTRUCTION LINES

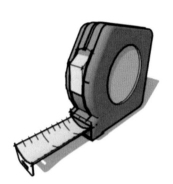

The Tape Measure is a multifunctional tool. Besides providing measurements it also **builds construction lines and points ("guides")**. Guides are used to help create geometry.

We're going to use the Tape Measure Tool to set up our first construction framework. We'll then use that framework to "hang" our drawing. Below is a primer on how the Tape Measure creates construction guides.

DRAWING GUIDES (LINES & POINTS) AT A SET DISTANCE

Try this exercise yourself

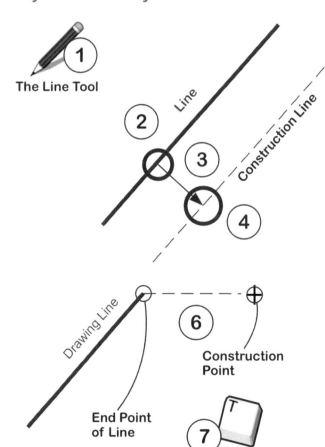

1. Draw a random line using the **Line Tool**.
2. Select the **Tape Measure Tool**, click on a **random** point on the line.
3. Move the mouse in the direction you want to place the (parallel) construction line.
4. Click to place the construction line at a random distance.
5. **Type a number** which represents the distance you want to give it and hit "enter". **The line is now at the distance from the line that you specified.**
6. Note that if you click on the **endpoint of a line**, you will get a **Construction Point, not a parallel line.** Again, move your mouse in the desired direction and type a distance from the point and hit "enter"
7. Useful Tip: Typing "T" will activate the Tape Measure (Default Shortcut.)

CONSTRUCTION DOCUMENTS USING SKETCHUP PRO 2020

DRAWING THE PLAN

1. Using the Tape Measure Tool, click on the edge "A" as indicated.
2. Move the cursor to the right and click a random distance away.
3. Type the number "682" (Don't click into any dialogue box to do this.)
4. Hit Enter.
5. The first construction line "B" is formed.

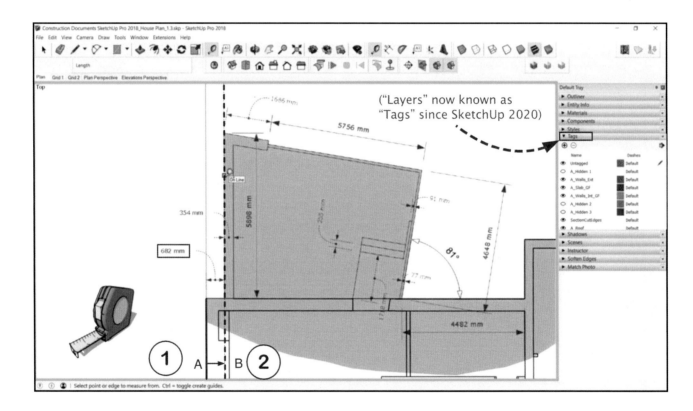

6. Click on the edge "C" as indicated.
7. Move the cursor upwards and click at a random distance.
8. Type the number "5898"
9. Hit Enter.
10. The construction line "D" is formed.

(Note that "Layers" (SketchUp 2019 and earlier) will here be referred to in this Manual as "Tags" (SketchUp 2020+)

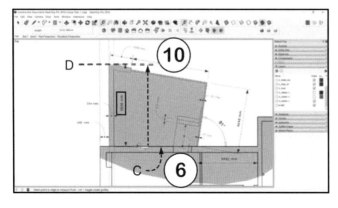

CONSTRUCTION DOCUMENTS USING SKETCHUP PRO 2020

DRAWING THE PLAN

Follow the procedure as outlined on the previous page to create the construction line "B" at a distance of "354" from "A"

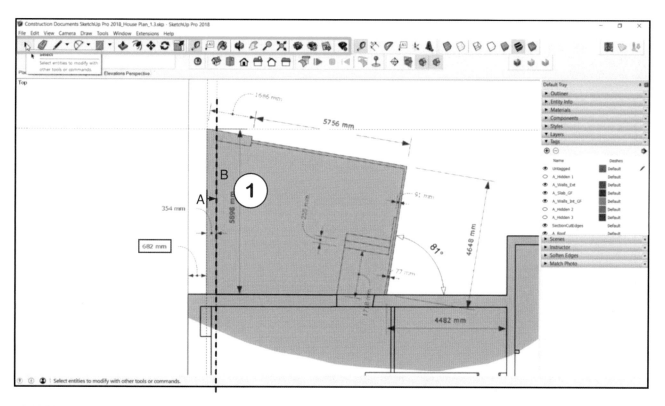

INFERENCING

As we draw lines or use other tools to move or rotate things, we need to be able to refer to lines, points, angles or planes that are already in place. This is what's known as **inferencing**. The diagrams below illustrate the various inferencing options.

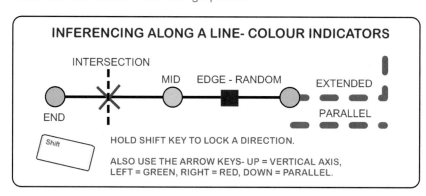

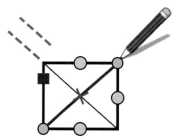

Drawing a line from one corner of an existing box to another corner is using inferencing to find the points of reference.

CONSTRUCTION DOCUMENTS USING SKETCHUP PRO 2020

DRAWING THE PLAN

Now we'll draw lines to start forming the angled corner. To form the edge "C" takes a few steps. First we'll form a construction point using a line.

1. Using the **Tape Measure Tool** click on the line "A" as indicated.
2. Using the procedure outlined previously, create a construction line "B" at a distance 4482 to the left of "A".

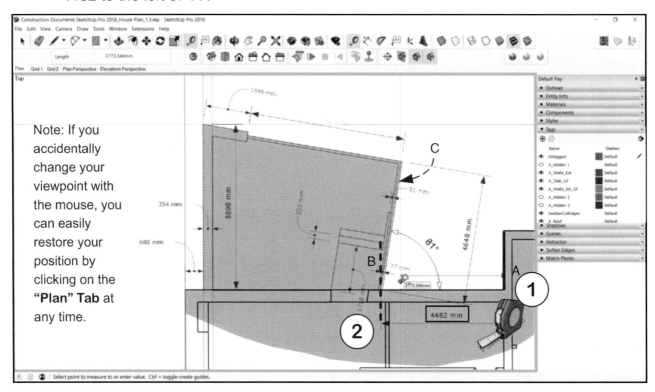

Note: If you accidentally change your viewpoint with the mouse, you can easily restore your position by clicking on the **"Plan" Tab** at any time.

The next tool we're going to use is the Protractor which also produces a construction line. It requires picking:

- A   A turning point (or "Fulcrum")
- B   Starting angle (0)
- C   A direction
- D   An angle of rotation

Sequence of Clicks using the protractor

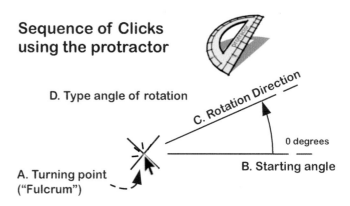

CONSTRUCTION DOCUMENTS USING SKETCHUP PRO 2020

DRAWING THE PLAN

Using the protractor tool:

We're now going to form a Construction Line using the **Protractor Tool**. The origin and the angle of rotation have to be established following the sequence below:

1. Click on the intersection point "Z" (between the construction line and the horizontal line.)
2. Pick a **zero** starting angle "A": Move the cursor to the right (We want to cursor to "stick" to the red axis.) Click.
3. Move the mouse upwards (Anticlockwise direction) and click a random angle.
4. Type "81" and hit Enter.
5. The construction line at "B" is formed.

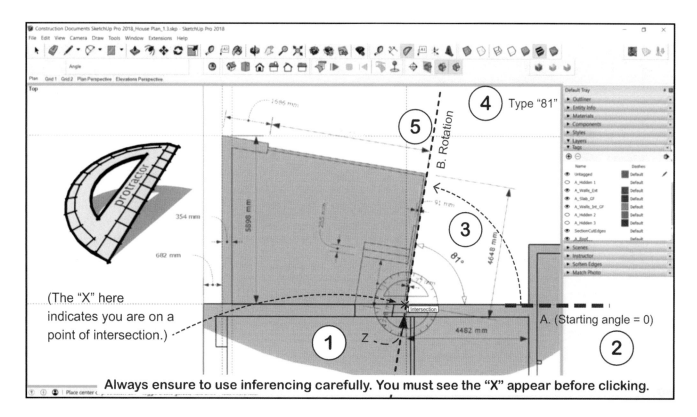

CONSTRUCTION DOCUMENTS USING SKETCHUP PRO 2020

DRAWING THE PLAN

1. Next: Using the **Line Tool**, draw a **line** starting from the point of intersection "X".
2. Move the cursor onto the construction line. Notice the red dot appearing- This indicates that your next point will be placed on that line.
3. Hold down the shift button- Notice that with the Shift button held down, wherever you place the cursor it is stuck on the construction line.

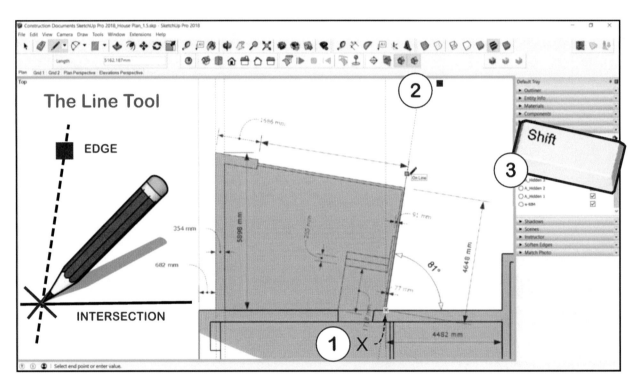

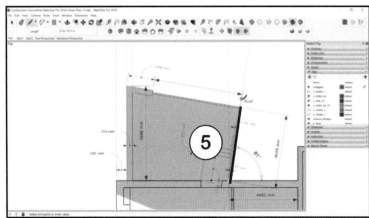

4. While holding down the Shift Button, type "**4648**" and hit enter. (Note: In doing this, don't click into the measurements dialogue- Just type the number.)
5. The line should appear as indicated.

DRAWING THE PLAN

To draw a line from the endpoint of the previous line at a perpendicular direction:

1. Start drawing a line from the end of the previous line- Let's call this "A". (Make sure to start from exactly the end point. See the green dot occurring below.)
2. Without clicking on your mouse button, **run the cursor along A**.
3. Notice when you move the cursor around that a magenta line appears perpendicular to A.

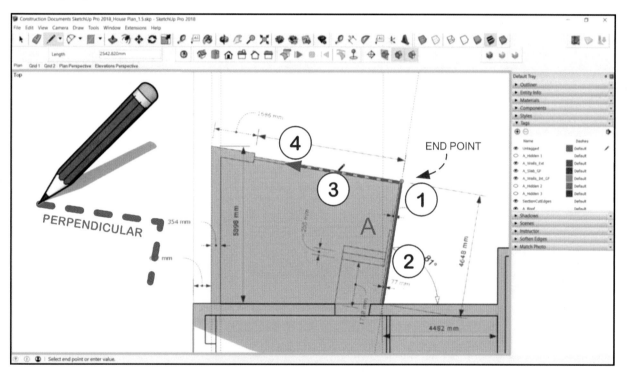

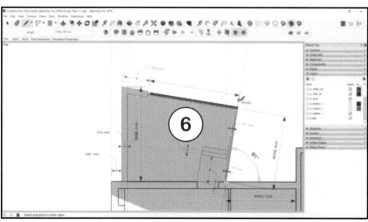

4 Move the cursor in the direction of the magenta line as illustrated. (Don't drag or click the mouse while doing this.)
5 With the line showing as magenta, type "**5000**" and hit Enter. (We'll complete it later.)
6 The resulting line should be as indicated.

DRAWING THE PLAN

Draw a line from the endpoint of the previous line at a perpendicular direction:

1. Start drawing a line from point X
2. Click on the intersection Y.
3. Start drawing a line from Point Y.
4. Float the cursor along line Z (without clicking) to pick up on it's direction.
5. Notice when you wiggle the cursor around that a **magenta line** appears parallel to Z.

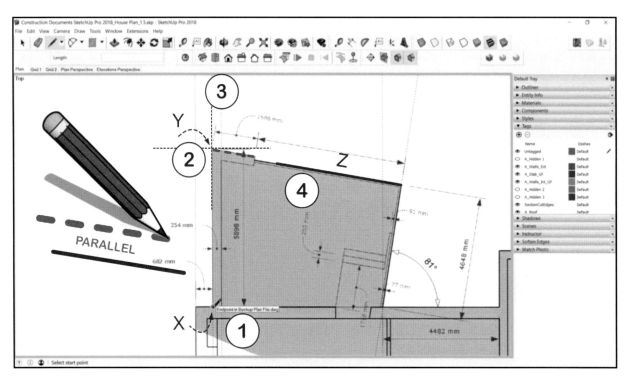

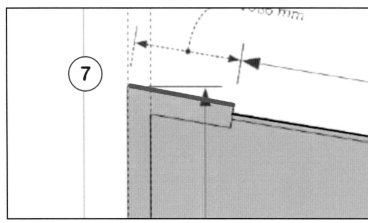

6. Whilst the magenta colour is showing, type "**1686**" and hit **Enter**.
7. The line should be as indicated.

Note: If you make a mistake, go back using CTRL + Z   (UNDO)

DRAWING THE PLAN

1. Using the Tape Measure Tool, click on a **random** point (not an end point) along **Line A** to create a **parallel construction line** at a distance of "354" from Line A.
2. Using the **Line Tool**, click on the point "X" to start a line from there.
3. Float the cursor along Line A to pick up it's direction.
4. Move the cursor around so that it turns magenta (perpendicular direction to Line A.) Draw a line B perpendicular to Line A which meets the construction line as illustrated.

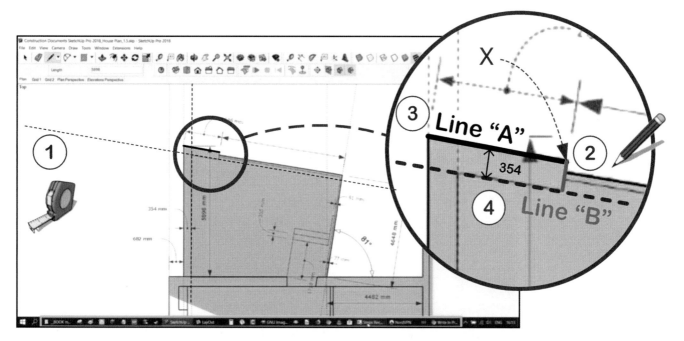

Finish drawing the shape as outlined below using the construction points you created.

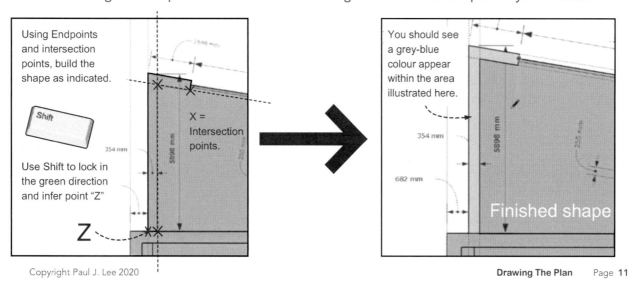

CONSTRUCTION DOCUMENTS USING SKETCHUP PRO 2020

DRAWING THE PLAN

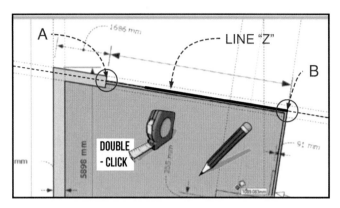

Finish the boundary shape.

1. Using the **Tape Measure Tool** double-click on Line "Z"
2. Using Select Tool, right-click on Line "Z" to delete it
3. Redraw Line "Z" from Point A (Endpoint) to Point B (Intersection)

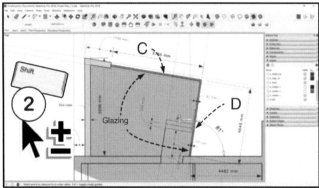

The next tools we're going to use are Select and Offset.

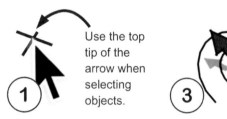

Use the top tip of the arrow when selecting objects.

SELECT TOOL   **OFFSET TOOL**

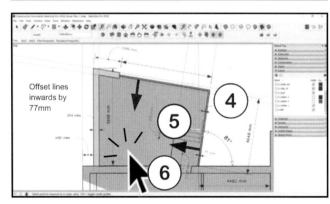

OFFSETTING LINES

Form inner lines to finish drawing the glazing around the perimeter.

1. Click on the **Select Tool**.
2. Holding down the Shift button, select lines C and D as indicated.
3. Click on the **Offset tool**.
4. Click on one or other of the selected lines.
5. Move the mouse inwards.
6. Click the mouse to select a random distance,
7. Type "77" and hit Enter.
8. The resulting lines should be as indicated.

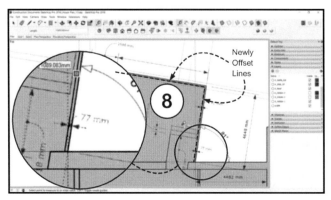

Drawing The Plan   Page **12**

DRAWING THE PLAN

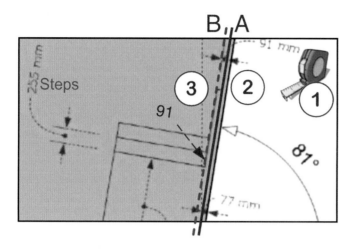

DRAWING THE STEPS

We're setting up construction lines to create a drawing of the steps. We'll first use the existing geometry of Line "A" to set up a parallel construction line at point "B".

1. Click on the **Tape Measure**.
2. Click on Line "A"
3. Click to the left (anywhere) of Line A and type a distance of "91" to create construction line "B"
4. To create the next construction line, click on Line "A" again.
5. Click on Intersection point "C". (This creates construction line "D".)
6. Select the **Line tool**.
7. Click on point "C" and float the cursor on construction line "D"
8. Type "1718" and hit Enter.
9. The resulting line should be as indicated.

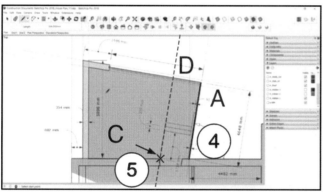

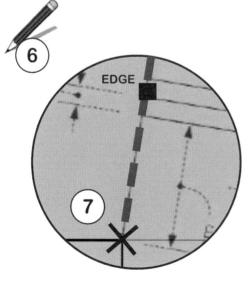

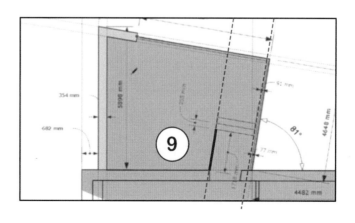

Copyright Paul J. Lee 2020

CONSTRUCTION DOCUMENTS USING SKETCHUP PRO 2020

DRAWING THE PLAN

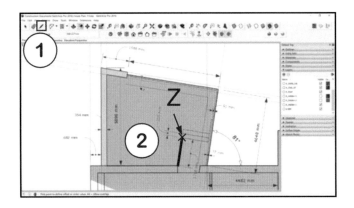

DRAWING THE STEPS

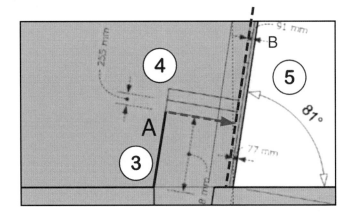

1. Select the **Line Tool**.
2. Click on the endpoint "Z".
3. Run the cursor along Line "A"
4. Move the cursor to the right until you see the perpendicular **magenta** inference line appear.
5. Move the cursor onto the construction line "B" and click.
6. The line formed should be as indicated.

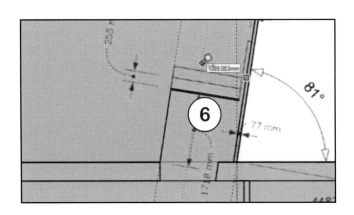

CONSTRUCTION DOCUMENTS USING SKETCHUP PRO 2020

DRAWING THE PLAN

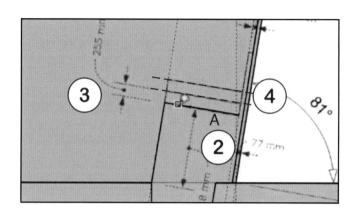

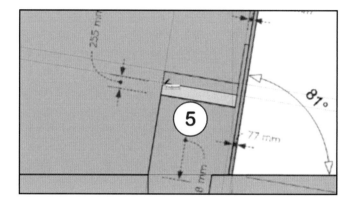

DRAWING THE STEPS

1. Select the **Tape Measure**.
2. Click on the line "A".
3. Click on any point above the line, type "255" and hit Enter.
4. Repeat to create a second construction line.
5. Start constructing the outlines of the steps using the methods previously outlined.
6. Draw the lines indicated in red. You should be able to complete the blue-grey region as indicated.

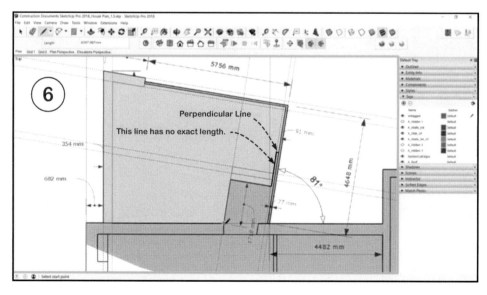

Faces are only formed when lines are properly joined end-to-end.

If there are any minute gaps- even so small you can't see them, then surfaces won't form.

CONSTRUCTION DOCUMENTS USING SKETCHUP PRO 2020

DRAWING THE PLAN

Handy Tip:
Drawing or editing a line that is "nearly" parallel to an axis:

To overcome the axis snap function, **zoom very close** into the area that you need to place the endpoint of your line. Zooming increases relative distance between elements to make positioning easier.

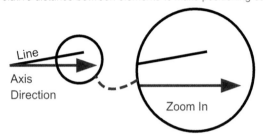

Finish the drawing.

To complete the grey-blue block as illustrated in below we first need to tidy up one section of the drawing.

1. Zooming closely into Corner "A" we see that Line "B" doesn't connect with the Intersect Line.
2. To extend Line "B" as required, we first need to create a construction line. Use the Tape Measure Tool to **double-click** on Line "B".
3. Using the Move Tool, click on the endpoint of "B" and then click on the intersection point.
4. Finish by drawing in the two line segments illustrated here.
5. Delete Lines C and D (These represent the line of the glazing but for the purposes of building our slab they must be eliminated.)

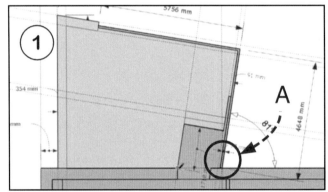

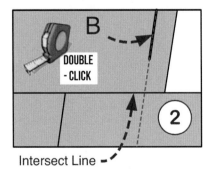

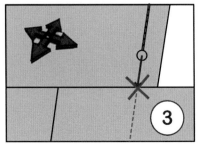

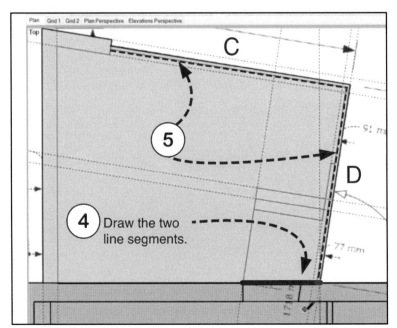

Copyright Paul J. Lee 2020

CONSTRUCTION DOCUMENTS USING SKETCHUP PRO 2020

DRAWING THE PLAN

Now we'll create a Local Component Collection for use in our model. This means we'll make the **SketchUp.Expert Collection** available so that we can insert SketchUp components from it.

1. Click on the **Components Dialogue** on the right hand side of your screen.
2. Click on the right-hand arrow button
3. Select "Open or Create A New Collection"
4. Navigate to the SketchUp.Expert folder (which we downloaded from the website.)
5. Double-click on the folder.
6. Now the Collection is available for you to select Components from it.
7. Click on the down arrow to see the full list of available collections.
8. Click into the SketchUp.Expert Collection to explore the components inside.

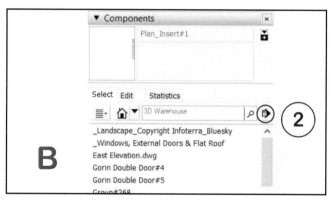
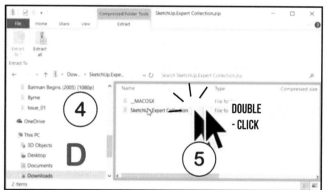

When you select the down arrow in the Components browser, you'll see the "**SketchUp.Expert Collection**" folder listed. Click on the folder to see the contents.

DRAWING THE PLAN

INSERTING THE DRAWING

Here we'll select the plan drawing from the Component List and insert it into the model. Then we'll explode this file to create regions such as: external walls - floor slab - internal walls.

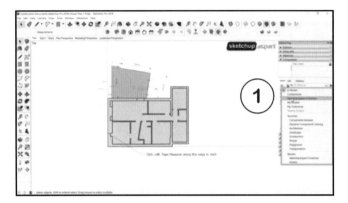

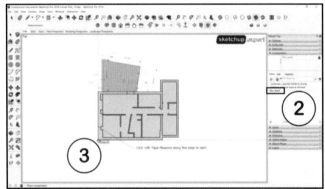

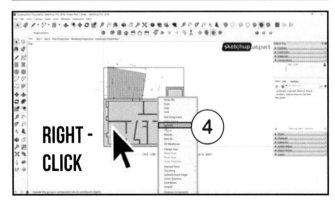

1. Select "SketchUp.Expert Collection" from the list.
2. Select the component called "_Plan_Insert" - This brings in the plan drawing.
3. Carefully select the zero point on the drawing plane as the location for the plan.
4. **Right-click on the plan and select "explode"**
5. Click on any one of the faces of the geometry.
6. Notice how it's broken into distinct regions defined by the boundary lines. These regions will form the basis for all your walls inside the model.

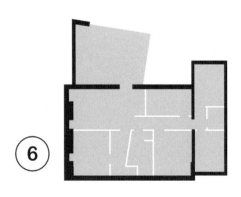

You can examine regions by observing the "blue dots" that appear when you click on a surface.

Regions

- External Walls
- Floor Slab
- Internal Walls

CONSTRUCTION DOCUMENTS USING SKETCHUP PRO 2020

CREATING GROUPS & LAYERS

Creation of Groups is fundamental to the structure of a SketchUp model. Here we start to create such a structure. The groups are organised in accordance with the regions outlined in the previous page. The first Group we're going to make is the Floor Slab.

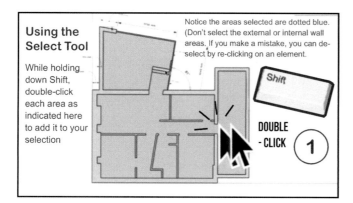

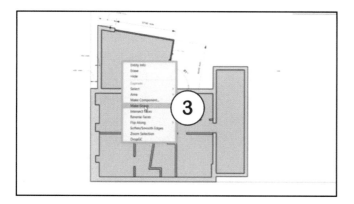

CREATING THE FLOOR SLAB GROUP

SELECT TOOL

1. Using the Select Tool (while holding Shift) select all areas associated with the Floor Slab (including the steps)
2. See the selected area is highlighted with blue dots.
3. When all the areas have been selected, right click **on** any part of this selection and select "**Make Group**"
4. Now the Floor Slab area has been grouped.

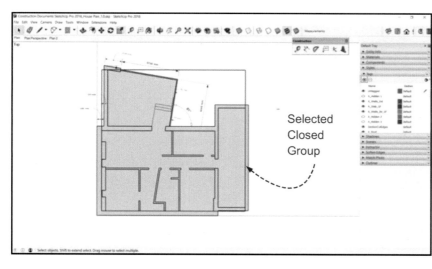

Closed Group:

Notice that a selected closed Group is signified by the following:

- Blue surround box
- Blue Object lines

When a group is closed it cannot be edited but it can be scaled and moved. To edit you have to open it. More on this later.

Copyright Paul J. Lee 2020

CONSTRUCTION DOCUMENTS USING SKETCHUP PRO 2020

CREATING GROUPS & LAYERS

The next step is to name the Floor Slab Group and put it on its own named Layer. This allows you to identify and locate the Group as you build your model.

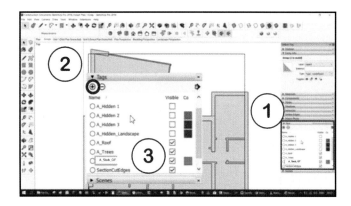

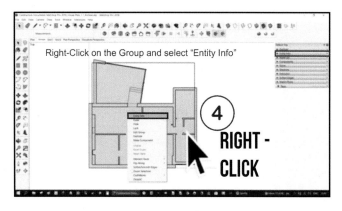

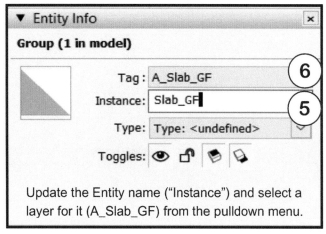

Update the Entity name ("Instance") and select a layer for it (A_Slab_GF) from the pulldown menu.

GROUP NAMING & LAYERS

1. Select the **Tags Dialogue**.
2. Click on the plus button.
3. A new dialogue opens to prompt you to name the new Tag. Call this tag "**A_Slab_GF**"
4. Next, expand the Entity Info dialogue by right-clicking on the group and selecting "**Entity Info**"
5. In Entity Info, name the Group Instance as "**Slab_GF**"
6. Place the Group onto the "A_Slab_GF" Tag using the pull-down menu.

This gives you full control of the Group for efficient editing and model organisation.

The importance of this will become clear later on.

Next we're going to repeat the process to create, name and organise the **External and Internal Walls Groups**.

Copyright Paul J. Lee 2020

CONSTRUCTION DOCUMENTS USING SKETCHUP PRO 2020

CREATING GROUPS & LAYERS

Now we're going to name the **External Walls Group** and **Ground Floor Internal Walls Group** and put these Groups on their own named Layers.

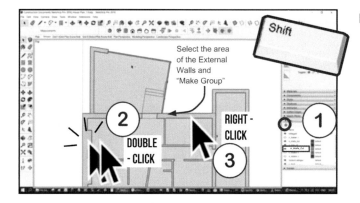

External Walls

1. Create a Tag called "**A_Walls_Ext**"
2. Select the external walls region as indicated. (Double-Click while holding Shift to select multiple areas.)
3. Right-click on the area and select "**Make Group**".
4. As in the previous example, with **Entity Info** give the Group a name. In this case it's "**Walls_Ext**".
5. With the Group selected, put it on the "**A_Walls_Ext**" Tag.

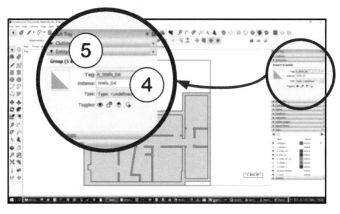

Internal Walls

6. Using the Select Tool (while holding Shift) select all areas associated with the Internal Walls.
7. Right-click and **Make Group**.
8. Create a Tag "**A_Walls_Int**" and name the Group as "**Walls_Int_GF**"

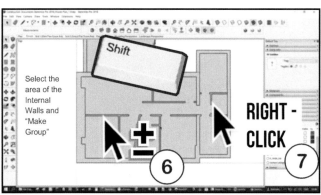
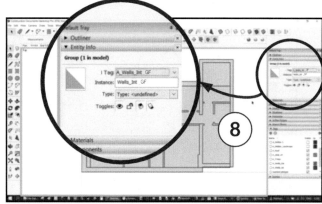

Copyright Paul J. Lee 2020   Creating Groups & Layers   Page **21**

CONSTRUCTION DOCUMENTS USING SKETCHUP PRO 2020

MODELLING THE HOUSE

1   Next we get to the 3D modelling part. We first of all start by selecting the **"Modelling Perspective" Scene Tab** at the top of the screen.

This scene has been set up to show a number of drawings all placed in the exact locations to correspond with the model we're going to build. These "prepared earlier" drawings allow us to concentrate on modelling only.

ELEVATING THE EXTERNAL WALLS

In the next part we're going to use the **Push-Pull Tool** to stretch the walls up to the height of the walls in the **Section Drawing**. First we need to open the Group for editing:

2   Using the select tool, right click on any part of the External Walls Group and select **Edit Group.**

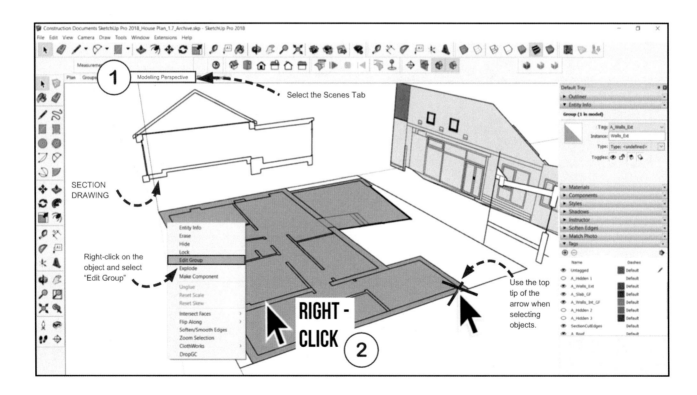

Copyright Paul J. Lee 2020   Modelling the House   Page **22**

CONSTRUCTION DOCUMENTS USING SKETCHUP PRO 2020

MODELLING THE HOUSE

PUSH-PULL TOOL + INFERENCING

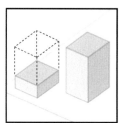 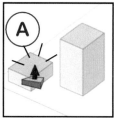 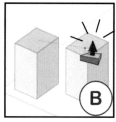 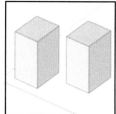

PUSH-PULL TOOL

Using the Push-Pull Tool to achieve height parity: Above are two (non-grouped) solids of differing heights. We want to make the smaller one as tall as it's neighbour. To do this, use the Push-Pull Tool. Float on the top surface of the solid as indicated and click on it (A). Move the cursor onto the top of the adjacent solid and click on it. (B). The two solids are now the same height.

ELEVATING THE EXTERNAL WALLS

1. Using the Push-Pull Tool, float over the wall region. (There are 3 distinct regions to elevate.)
2. Click on the surface to start elevating.
3. When you start to elevate the walls, move the cursor to the corresponding location on the section drawing.
4. Click on the corresponding location.
5. Your walls should match the image below.

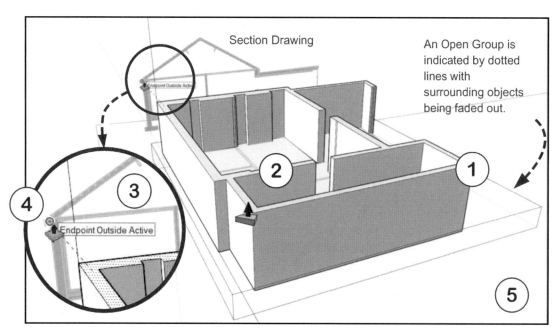

CONSTRUCTION DOCUMENTS USING SKETCHUP PRO 2020

MODELLING THE HOUSE

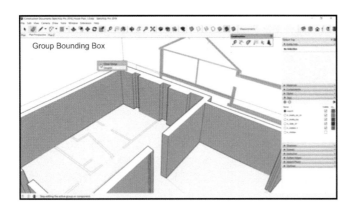

CLOSE THE WALLS_EXT GROUP

Right-Click into an area outside of the bounding box and select "Close Group"

SELECT TOOL   **MOVE TOOL**

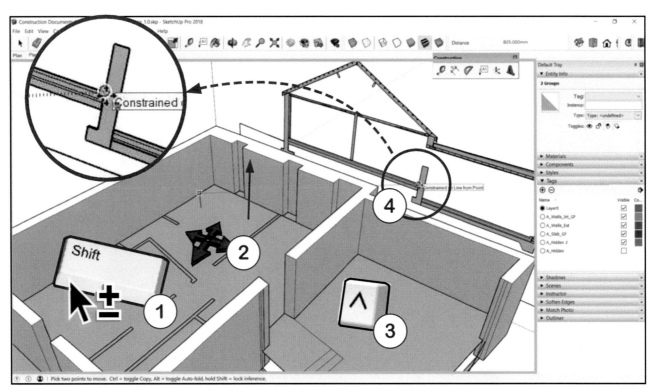

REPOSITION THE FLOOR SLAB
We're going to move the Floor Slab upwards to the level indicated on the Section Drawing:

1. Using the **Select Tool**, with a single click select the **Floor Slab and the Internal Walls Groups**.
2. Select the **Move Tool** and click once on the Slab to start moving it upwards
3. While moving the slab, **hit the Up Arrow on your keyboard** to **lock in the Blue Direction**.
4. Click on the Adjacent Line representing the top level of the Floor Slab.
5. **When you're done, don't forget to close the Group.** (Right-click on blank space and select "Close Group")

CONSTRUCTION DOCUMENTS USING SKETCHUP PRO 2020

"POWERFUL SHORTCUT No. 1": HIDE

SketchUp has the ability to instantly turn on and off visibility for geometry outside of an Open Group. This is hugely advantageous and greatly increases modelling speed.

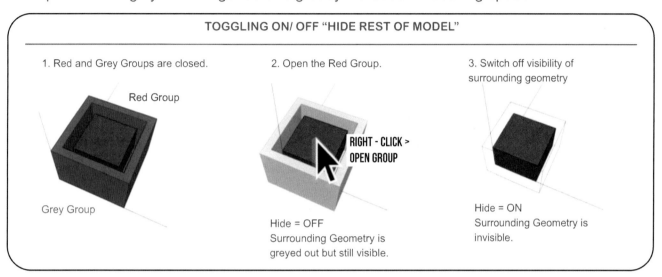

TOGGLING ON/ OFF "HIDE REST OF MODEL"

1. Red and Grey Groups are closed. (Red Group, Grey Group)
2. Open the Red Group. RIGHT - CLICK > OPEN GROUP. Hide = OFF Surrounding Geometry is greyed out but still visible.
3. Switch off visibility of surrounding geometry. Hide = ON Surrounding Geometry is invisible.

Let's create a shortcut for this function.

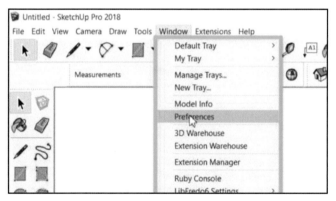

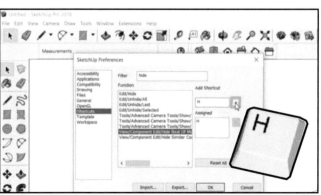

1. Select **Window > Preferences > Shortcuts**. (SketchUp > Preferences > Shortcuts for Mac)
2. Type "**hide**" into the filter search.
3. Click on the selection called: **View> Component Edit> Hide Rest of Model.**
5. Type "**H**" into the "Add Shortcut" Dialogue and click on the "+" button.
6. You will be asked if it's OK to override an existing function. Select "Yes".

Now the shortcut is ready. Test the effectiveness of the shortcut by opening any group in your model. Hit the "H" key and see how the surrounding geometry disappears. To make it reappear just hit "H" again. We're going to use this shortcut in our modelling from here on.

CONSTRUCTION DOCUMENTS USING SKETCHUP PRO 2020

MODELLING THE HOUSE

EDITING THE SLAB

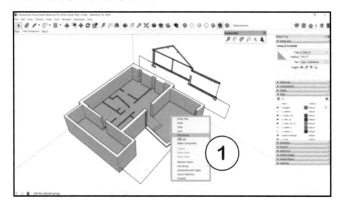

1. Right-Click on the Slab Group and select "**Edit Group**"
2. Activate the "H" Shortcut to Hide Rest of Model
3. "Repair" the slab (fill in the empty parts) by redrawing the edges as illustrated.
4. Delete the lines.
5. Using the Push-Pull Tool Click on the isolated area of floor as indicated here.
6. (Can't see the section drawing? Press "H" again.)
7. With the section drawing visible, move the cursor over to the lower floor level.

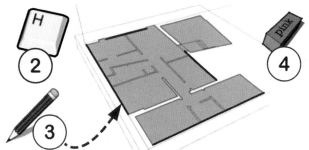

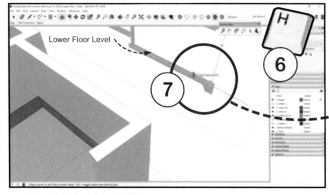

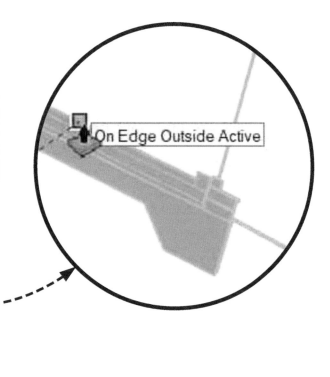

MODELLING THE HOUSE

EDITING THE SLAB

Here we clean up the slab group and create the three steps.
(Reminder: Make sure the Slab Group is open. Otherwise you can't edit it.)

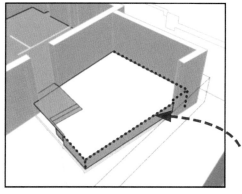

Delete top and side edges

1   With the Slab Group open, using the Eraser Tool delete the edges of the surfaces indicated here.

ERASER TOOL   **SELECT TOOL**

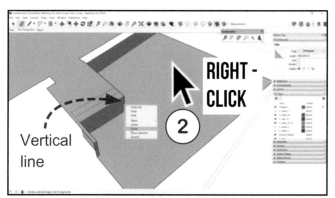

Vertical line

2   Now right-click on the vertical line and select "Divide"

3   You'll notice that red squares appear on the line. They increase and decrease as you slide the cursor up and down. When you arrive at **3 divisions**, click to accept.

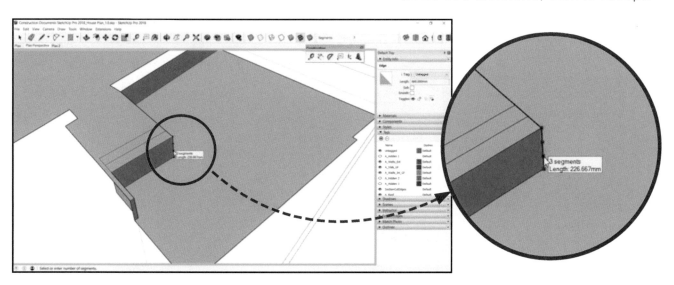

MODELLING THE HOUSE

EDITING THE SLAB- Finishing off the Steps

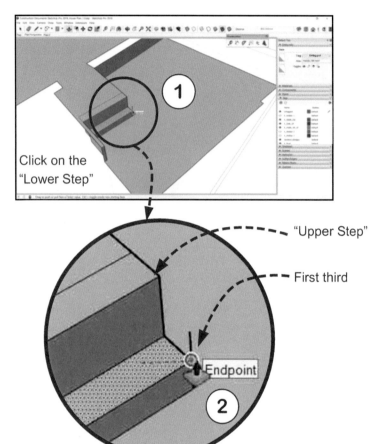

PUSH-PULL TOOL

1. Using the **Push-Pull Tool**, click on the surface of the "**Lower Step**".
2. Move the cursor until it meets the top of the lowest line division "**Endpoint**". Click to accept.
3. Click on the surface of the "**Upper Step**" and repeat the process.
4. With the steps completed, **close the Group.**

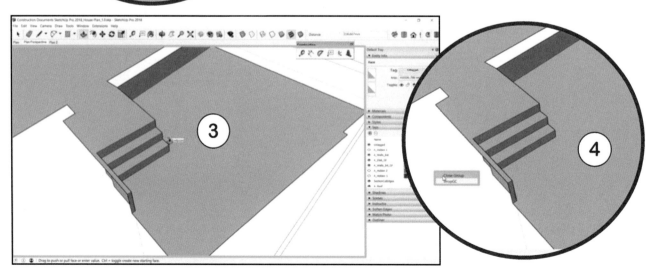

CONSTRUCTION DOCUMENTS USING SKETCHUP PRO 2020

STRUCTURING YOUR MODEL

Knowing how a model is structured means that we can work efficiently. To do this, we have a dialog called "Outliner" which lists all the groups, components and "Section Cuts"* inside it. Outliner describes the entire model in a "Tree/ Branch" structure. It also tells us what group we're currently editing. Any Group or Component can contain "Sub-Groups" or "Sub-Components".

"OUTLINER" EXAMINES THE STRUCTURE OF YOUR MODEL AND ALLOWS YOU TO CONTROL IT.

■ GROUP   ⌖ SECTION CUT   ⌂ SKETCHUP FILE (HOME)
▪▪ COMPONENT   (CLOSES ALL GROUPS/ COMPONENTS)

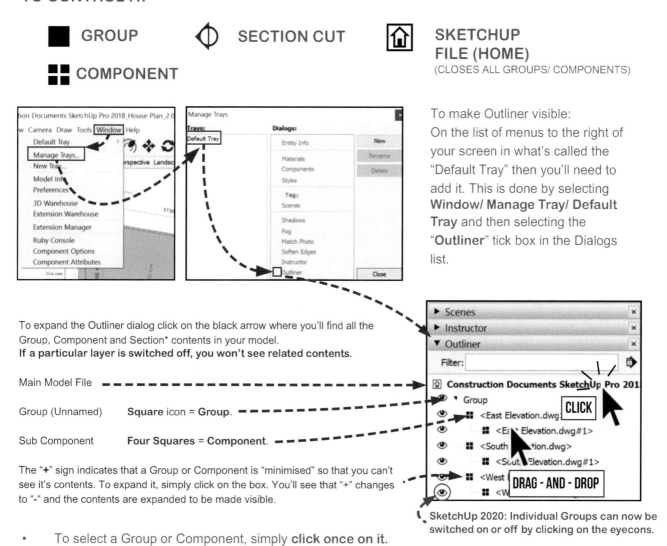

To make Outliner visible:
On the list of menus to the right of your screen in what's called the "Default Tray" then you'll need to add it. This is done by selecting **Window/ Manage Tray/ Default Tray** and then selecting the "**Outliner**" tick box in the Dialogs list.

To expand the Outliner dialog click on the black arrow where you'll find all the Group, Component and Section* contents in your model.
If a particular layer is switched off, you won't see related contents.

Main Model File

Group (Unnamed)   **Square** icon = Group.

Sub Component   **Four Squares = Component.**

The "+" sign indicates that a Group or Component is "minimised" so that you can't see it's contents. To expand it, simply click on the box. You'll see that "+" changes to "-" and the contents are expanded to be made visible.

SketchUp 2020: Individual Groups can now be switched on or off by clicking on the eyecons.

- To select a Group or Component, simply **click once on it.**
- To edit the component itself, **double-click on it, or right click to select "edit"**
- You can also **drag and drop any Group or Component** into any other Group or Component.

CONSTRUCTION DOCUMENTS USING SKETCHUP PRO 2020

STRUCTURING YOUR MODEL

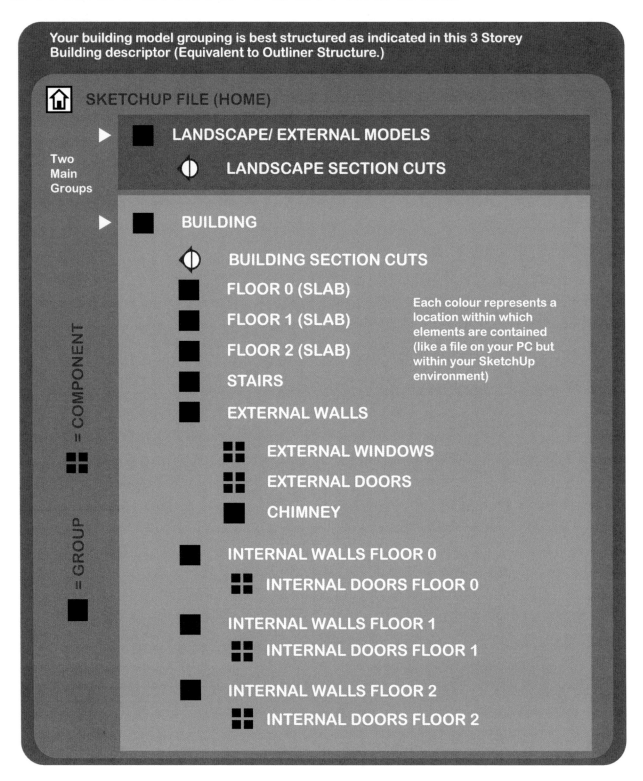

MODELLING THE HOUSE

EDITING THE INTERNAL WALLS

Now that we've done work on the External Walls and Slab Groups we're going to tackle the inside walls.

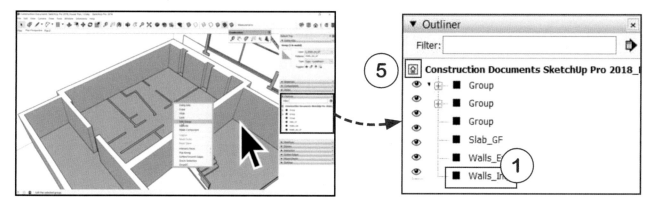

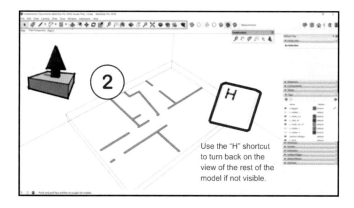

1. To open the internal Walls Group, expand Outliner and select "Walls_Int_GF" from the list. (Double-click to open)
2. Select the Push-Pull Tool and click on any one of the wall surfaces.
3. Move the cursor level with the top of the external walls.
4. Repeat this process with all the internal walls until they are complete.
5. Close all Groups- You can either click on the Home icon in Outliner or Right Click on the background and select Close.

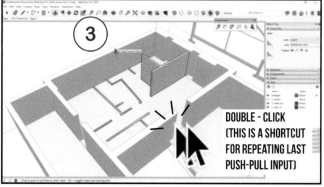

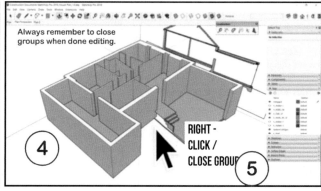

CONSTRUCTION DOCUMENTS USING SKETCHUP PRO 2020

MODELLING THE HOUSE

EDITING THE SLAB- Adding thickness.

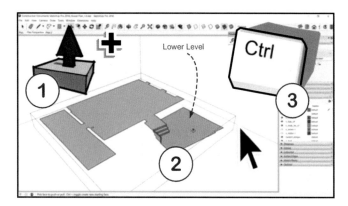

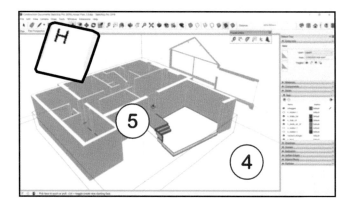

Open the **Slab_GF Group**

1. Select the Push Pull Tool.
2. Click on the lower slab surface as illustrated and move the mouse downwards.
3. If the surface doesn't produce a thickness, hit the CTRL key (or the alt key for Mac users) to add a new surface. The "+" sign appears next to the Push Pull Icon.
4. Move the cursor over to the section drawing to infer the corresponding depth of the slab.
5. Repeat this process for the higher floor level.

You may have to fix some geometry that contains holes. This is done by simply redrawing lines that "support" the geometry. Normally this fixes the surfaces.

Sometimes geometry may not behave in a predictable way- Geometry may not form or it may not allow you to edit it. The next page contains some pointers on dealing with this.

ORIENTING THE FACES

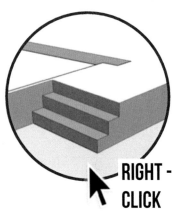
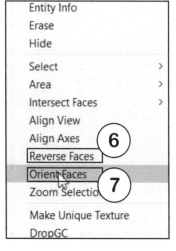

Surfaces can be formed "inside out" (blue-grey)

6. To reverse any surface, right-click on it and select "Reverse Faces". You can select multiple surfaces at a time- (CRTL + A selects all surfaces at once) or alt + A for Mac.
7. To orient all surfaces to match, right-click on a face and select "Orient Faces".

Copyright Paul J. Lee 2020

CONSTRUCTION DOCUMENTS USING SKETCHUP PRO 2020

FIXING GEOMETRY

MISSING SURFACE ON A BOX

When we have a missing face such as this, the normal way to fix it is to re-draw an edge.

When this doesn't work, there is some problem with the geometry. The list of potential geometry problems are:

- Broken edges (multiple end points)
- Hidden/ Buried lines
- Off-axis lines
- Non-coplanar faces

It's difficult to figure out which problem you're encountering.

Redrawing any edge will normally "heal" the object.

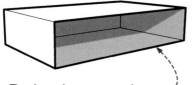

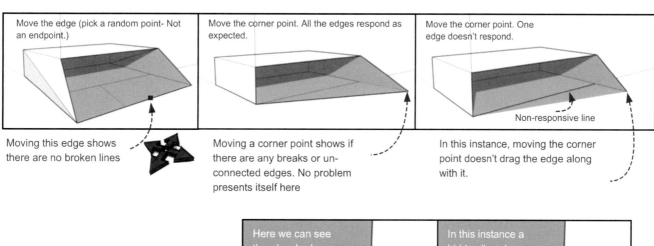

Move the edge (pick a random point- Not an endpoint.)

Moving this edge shows there are no broken lines

Move the corner point. All the edges respond as expected.

Moving a corner point shows if there are any breaks or un-connected edges. No problem presents itself here

Move the corner point. One edge doesn't respond.

Non-responsive line

In this instance, moving the corner point doesn't drag the edge along with it.

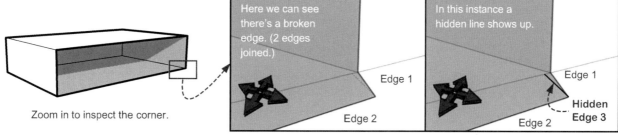

Zoom in to inspect the corner.

Here we can see there's a broken edge. (2 edges joined.)

Edge 1
Edge 2

In this instance a hidden line shows up.

Edge 1
Hidden Edge 3
Edge 2

To fix the surface, delete the faulty edges and redraw.

CONSTRUCTION DOCUMENTS USING SKETCHUP PRO 2020

MODELLING THE HOUSE

MAKING THE ROOF

To make the hip roof, we're going to create a profile of "half" the roof and then extrude this profile around in a "horseshoe" shape to complete the hip roof shape.

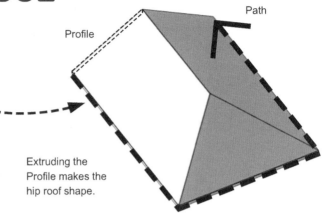

Extruding the Profile makes the hip roof shape.

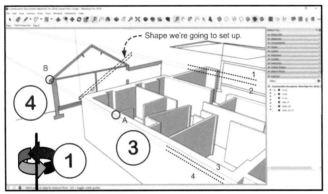

SET UP CONSTRUCTION LINES

1. Orbit into a good position like that illustrated
2. Select the **Tape Measure Tool**.
3. Click on a random point along the long edge of the wall (A)
4. Click on a node point on the section drawing as indicated (B)
5. Repeat this process to create four parallel construction lines as indicated.

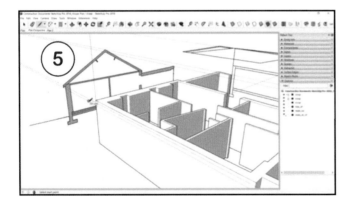

DRAW THE ROOF PROFILE

6. Using the **Line Tool**, let the cursor rest on a Construction line.
7. Holding down the **Shift Key**, move the cursor to the **back edge** of the external wall.
8. Click the mouse in this location to begin drawing the profile shape.
9. Repeat the process on each construction line to complete.

Go to the next page >>>

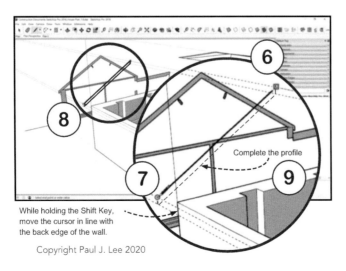

While holding the Shift Key, move the cursor in line with the back edge of the wall.

CONSTRUCTION DOCUMENTS USING SKETCHUP PRO 2020

MODELLING THE HOUSE

CREATE THE MAIN ROOF USING THE FOLLOW ME TOOL

THE FOLLOW-ME TOOL

A. Select an extrusion path.

B. Click on the Follow-Me Tool

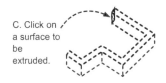
C. Click on a surface to be extruded.

D. Finished Object

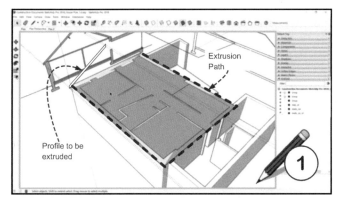

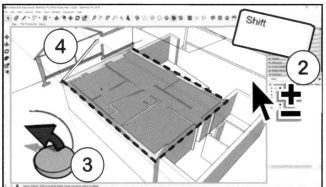

DRAWING THE EXTRUSION PATH

1. Draw a rectangle around the outer edges of the main part of the house. (Make sure lines are precisely located on each corner.)
2. Using the Select Tool select the extrusion path- Hold down the **Shift Key**, select **the three lines** as indicated.
3. Select the **Follow Me Tool**.
4. Click on the profile.
5. **The roof shape should appear instantly.**
6. **Triple-click** on the roof geometry to select the entire roof.
7. If the roof appears blue-grey, right click and select "Reverse Faces"- If the surfaces are white then leave as-is.
8. Right-Click and select "Make Group"
9. In Outliner right-Click on the new Group
10. Select "Rename" on the Menu
11. Type "**Roof 01**" as the name for the Group.

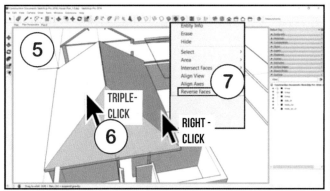

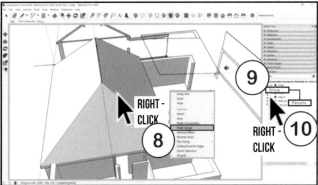

Copyright Paul J. Lee 2020 — Modelling the House — Page 35

CONSTRUCTION DOCUMENTS USING SKETCHUP PRO 2020

MODELLING THE HOUSE

Create a Tag called "**A_Roof**" on which we'll place the main roof and side roof Groups.

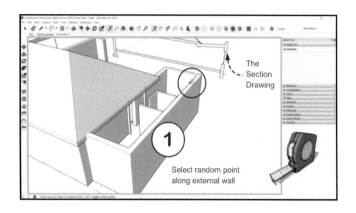

CREATE THE SIDE EXTENSION ROOF

First we'll create the roof profile to be extruded using construction lines coordinated with the Section Drawing.
Then we'll extrude this shape to create the roof.

THE CONSTRUCTION FRAME

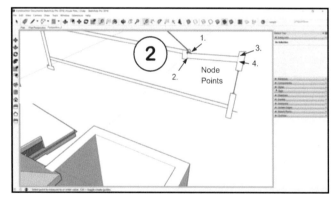

1. Using the Tape Measure Tool, click on a random point along the external wall.
2. Place construction lines by clicking on node points.
3. Repeat the process until the four construction lines are in place.

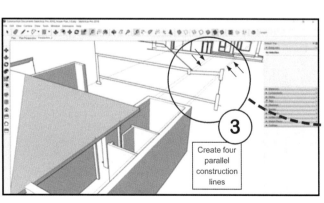

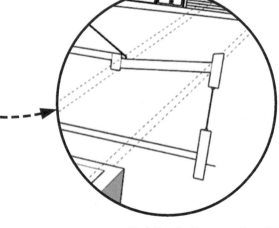

MODELLING THE HOUSE

CREATE THE SIDE EXTENSION ROOF

Here we'll draw the Profile in an isolated location and extrude it to create the roof.

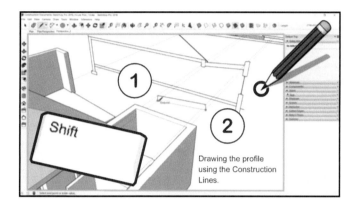

Drawing the profile using the Construction Lines.

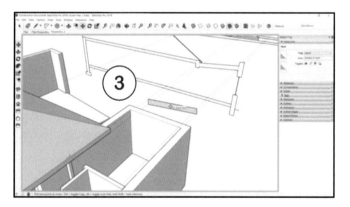

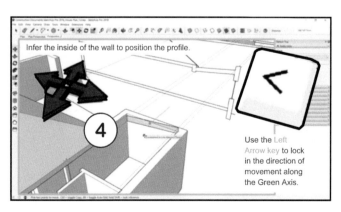

Infer the inside of the wall to position the profile.

Use the Left Arrow key to lock in the direction of movement along the Green Axis.

USING THE LINE TOOL TO CREATE THE PROFILE

1. The first point that you choose on the Construction Line is random- The next three must coordinate with the first.
2. To place the next point of the line rest the cursor on a Construction Line
3. While holding Shift, click **on the previous endpoint**. Repeat this process to complete the profile shape. If properly formed, the shape will have an internal surface.
4. Move the profile into place against the inside of the Side Extension. Use the **Move Tool** to select and move the profile (shape) along the Green Axis. **Use the Left Arrow Key on your keyboard to lock in the direction.**

CONSTRUCTION DOCUMENTS USING SKETCHUP PRO 2020

MODELLING THE HOUSE

FINISH THE SIDE EXTENSION ROOF AND PARAPET WALLS

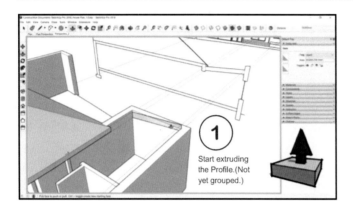

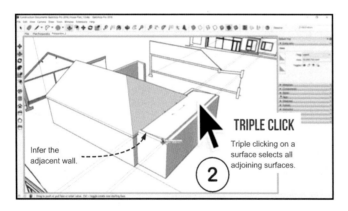

1. Use the **Push Pull Tool** to extrude the profile along the length of the extension as illustrated.
2. To create a Group for the Roof, using the select tool **triple click** on the roof surface to select it, right-click on the selection and select "**Make Group**"
3. Right-click and Select "**Entity Info**" Name this Group "**Roof_02**"
4. Place the Group on "A_Roof" layer
5. To edit the parapet, Right-Click on the **Walls_Ext** Group and select "**Edit Group**".
6. Use the **Push Pull Tool** to raise the height of the wall to coincide with the profile drawing as illustrated.

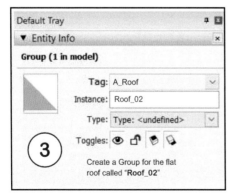

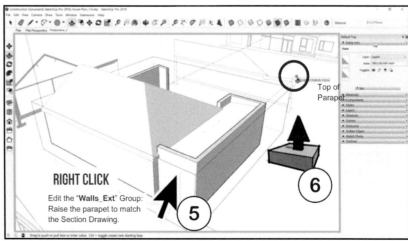

CONSTRUCTION DOCUMENTS USING SKETCHUP PRO 2020

MODELLING THE HOUSE

FINISHING THE PARAPET WALLS

Use the Tape Measure, Line and Push Pull Tools to shape the top of the Parapet.

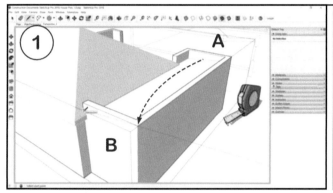

Using the Tape Measure Tool, click a random point on Edge "A" and place it on the corner point "B"

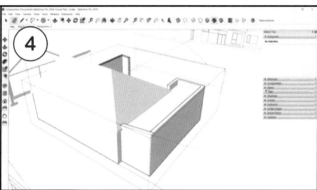

This is what the result should look like so far.

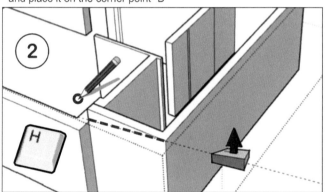

Draw a Line over the Construction Line and Push Pull to other face.

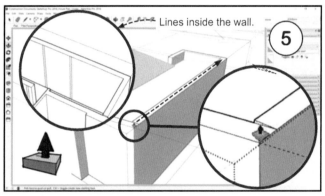

Push Pull the small triangle that was left over. Note that as you push pull across, the geometry gets "stuck". This is due to the lines that exist inside the wall (See illustration) To overcome this, keep push-pulling incrementally until you reach the end.

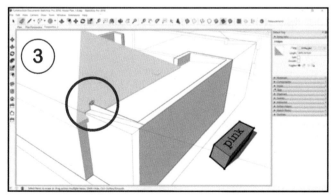

Delete stray lines using Eraser Tool.

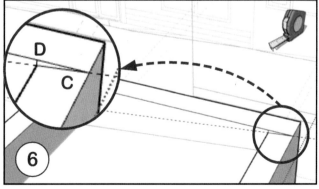

Using Tape Measure, double-click on Line C.
Delete Line D.

MODELLING THE HOUSE

FINISHING THE PARAPET WALLS

Use the Tape Measure, Line and Push Pull Tools to shape the top of the Parapet.

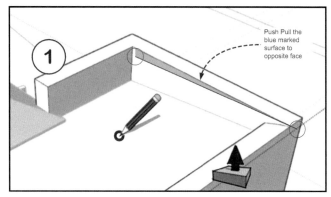

Draw a line along the construction line and Push Pull to face.

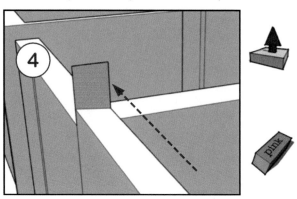

Push-pull shape to the end of the wall. Delete unnecessary lines (if any)

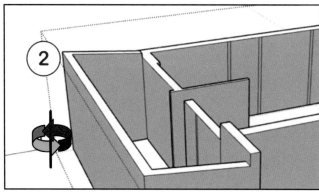

Orbit around to the other side of the wall

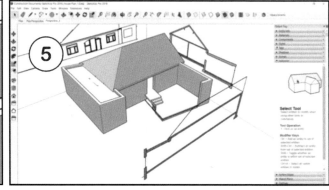

This is what the end result should look like.

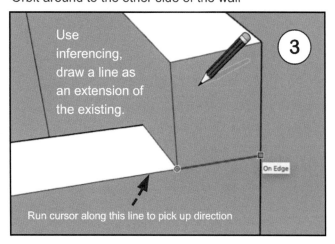

CONSTRUCTION DOCUMENTS USING SKETCHUP PRO 2020

MODELLING THE HOUSE

FINISHING THE PARAPET WALLS

Use the Tape Measure, Line and Push Pull Tools to shape the top of the Parapet.

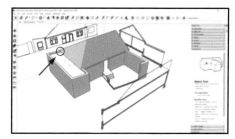

Parapet detail at front side of the house

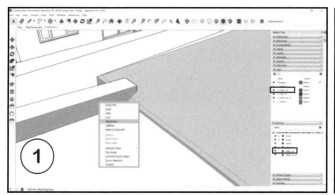

1. Open the "Walls_Ext" group using **Outliner** or Right-Click-Edit Group.

2. Using Tape Measure select a random point on the end line and create a construction line (say 200mm away from edge)

Use the Tape Measure to create a construction line.

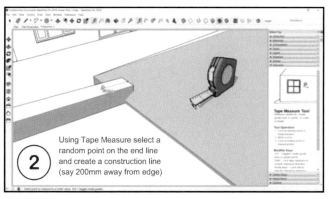

3. A. Draw a line from the intersection of the construction line and the edge of the parapet.
B. Lock into the blue (vertical axis using the up arrow on your keyboard.) and click on the roof surface.

UP ARROW

Draw a line along the blue axis to start making the "cut-out" shape.

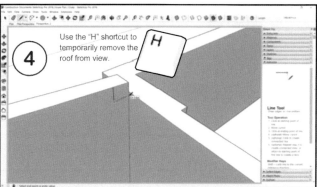

4. Use the "H" shortcut to temporarily remove the roof from view.

Draw a line along the red axis to finish shape

5. Push-Pull the remaining piece of wall to complete the parapet.

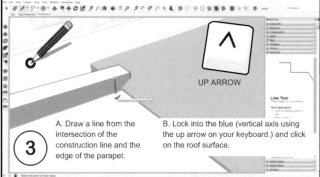

6. To finish the detail, push-pull the horizontal surface to align with the underside of the soffit. (Orbit around to view.)

Soffit line

This is what the end result should look like.

Copyright Paul J. Lee 2020   Modelling the House   Page 41

CONSTRUCTION DOCUMENTS USING SKETCHUP PRO 2020

MODELLING THE HOUSE

PLACING OPENINGS IN THE WALLS

Use the Rectangle tool to draw openings that correspond to the elevation drawings.

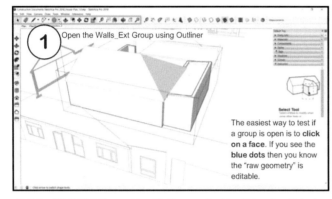
Edit the Wall_Ext Group. Use Outliner or double-click using Select.

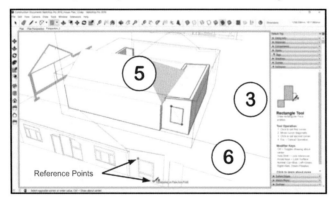

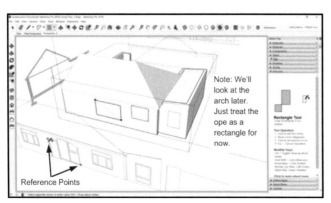
Finish drawing all the opes using the above method.

RECTANGLE TOOL

Open the **Walls_Ext** Group.

To draw a rectangle onto the wall:

1. Ensure the correct Group is open. (Test by clicking on a face to see if blue dots show up.)
2. Select the Rectangle Tool.
3. Float the cursor onto the surface in which you wish to create the opening.
4. Hold down the Shift button to lock the cursor onto the surface. Note the magenta colour appears which indicates the "locking" effect.

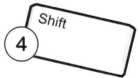

5. Float the cursor onto a reference point.
6. Click on the reference point.
7. Click on the opposite point
8. The rectangle is now formed.
9. Finish all the opes using the same method.

Copyright Paul J. Lee 2020

CONSTRUCTION DOCUMENTS USING SKETCHUP PRO 2020

MODELLING THE HOUSE

FORMING OPENINGS IN THE WALLS

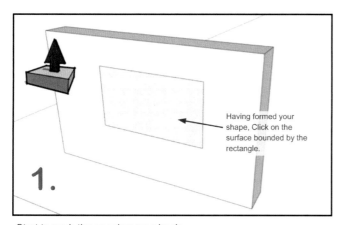

Having formed your shape, Click on the surface bounded by the rectangle.

Start to push the opening area back

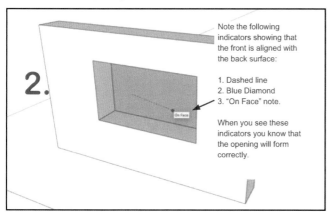

Note the following indicators showing that the front is aligned with the back surface:

1. Dashed line
2. Blue Diamond
3. "On Face" note.

When you see these indicators you know that the opening will form correctly.

When the front surface meets the back surface they cancel out.

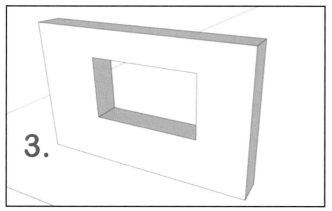

Opening formed.

Note: We'll look at the arch later. Just treat the ope as a rectangle for now.

PUSH-PULL TOOL

Finish creating all the openings:

1. Ensure the correct Group is open.
2. Select the Push Pull Tool.
3. Click on the surface inside the rectangle and push this surface to the adjacent one.
4. Finish all the opes to the front and the rear in a similar way, using the drawings as a guide.

If you aren't operating inside the group which contains the geometry you wish to work on, none of the drawing and editing tools will work- (as in: splitting faces, push-pull, offset etc.)

CONSTRUCTION DOCUMENTS USING SKETCHUP PRO 2020

CREATE "POWERFUL SHORTCUT No. 2"

SketchUp allows you to examine which lines are conforming to the current axes. **Model faces need to be precisely parallel for faces and lines to interact correctly.** Examining lines using the following method is a convenient and quick way of ensuring that lines and surfaces are parallel with each other.

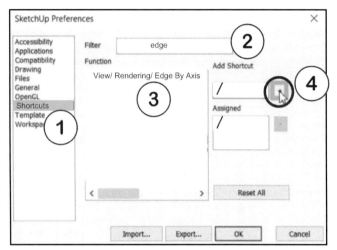

Set up the shortcut through Preferences (On a Mac: SketchUp/ Preferences)

1. Select **Window/ Preferences/ Shortcuts**. (SketchUp > Preferences > Shortcuts for Mac)
2. Type "**edge**" into the filter search.
3. Click on the selection called: **View/ Rendering/ Edge/ By Axis**
4. Type "**/**" into the "Add Shortcut" Dialogue and click on the "**+**" button.
5. If you're asked if it's OK to override an existing function. Select "Yes".

Now the shortcut is ready. Test the effectiveness of the shortcut by hitting the the "/" key and see how the axis-aligned lines change colours.

Lines parallel to red axis turn red, green axis green, blue etc.

Lines that are not aligned to an axis will show black.

To make the lines return to their normal appearance, just hit "/" again.

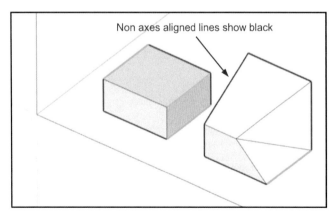

Pressing the "/" shortcut shows up all lines that are parallel to an axis. Pressing it again turns the lines back to normal.

Use this shortcut to check your model especially where you're finding geometry tricky to deal with.

MODELLING THE HOUSE

PLACING OPENINGS IN THE WALLS

Using the 2 point Arc Tool, draw the arc as illustrated below, and complete the window ope.

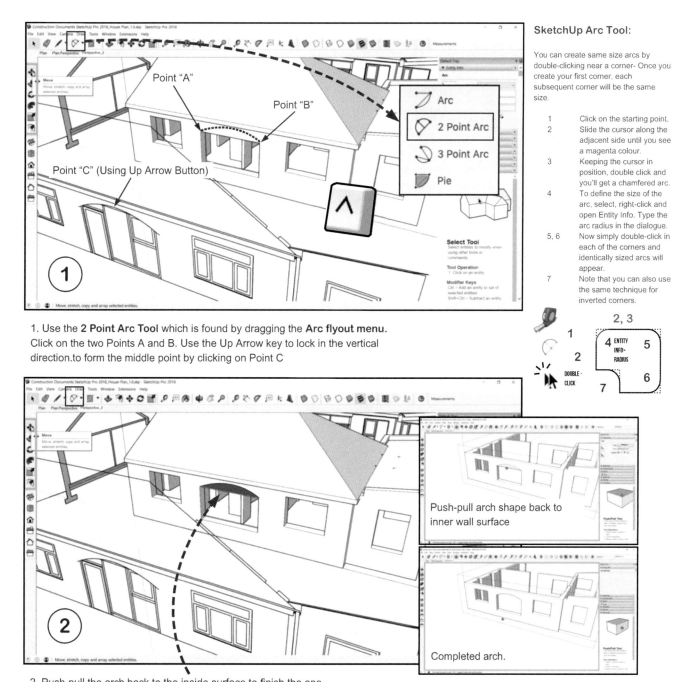

SketchUp Arc Tool:

You can create same size arcs by double-clicking near a corner- Once you create your first corner, each subsequent corner will be the same size.

1. Click on the starting point.
2. Slide the cursor along the adjacent side until you see a magenta colour.
3. Keeping the cursor in position, double click and you'll get a chamfered arc.
4. To define the size of the arc, select, right-click and open Entity Info. Type the arc radius in the dialogue.
5, 6. Now simply double-click in each of the corners and identically sized arcs will appear.
7. Note that you can also use the same technique for inverted corners.

1. Use the **2 Point Arc Tool** which is found by dragging the **Arc flyout menu**. Click on the two Points A and B. Use the Up Arrow key to lock in the vertical direction.to form the middle point by clicking on Point C

2. Push-pull the arch back to the inside surface to finish the ope.

CONSTRUCTION DOCUMENTS USING SKETCHUP PRO 2020

MODELLING THE HOUSE

CREATING THE WINDOW COMPONENT

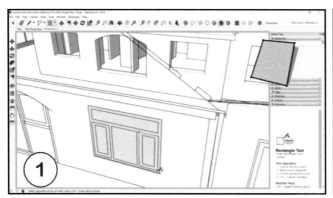

Draw a series of rectangles onto the elevation drawing to describe the parts of the window. Redraw all the rectangles to create different regions.

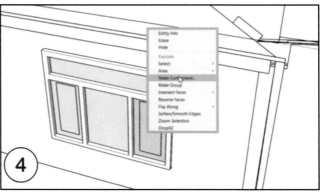

Right Click on the selected parts and choose "Make Component"

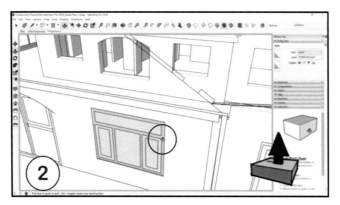

Push - pull the window frame outwards by 85mm to start creating the body of the window

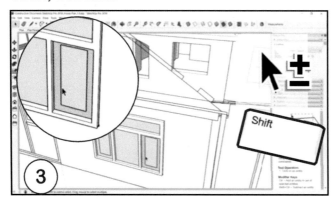

Using the Select Tool double-click the inner frame surface to select the sash element. Hold down Shift and double-click on the inner surface.

Name the element as "Window_Leaf" Make sure "Replace selection with component" is selected, and click "Create".

CONSTRUCTION DOCUMENTS USING SKETCHUP PRO 2020

MODELLING THE HOUSE

CREATING THE WINDOW COMPONENT

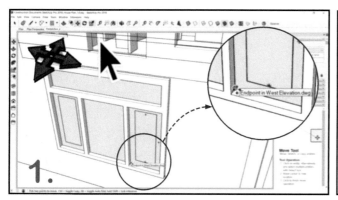

1. Select the closed Window Leaf Component and, **selecting the bottom left corner of the leaf frame** move towards the front edge of the window frame.

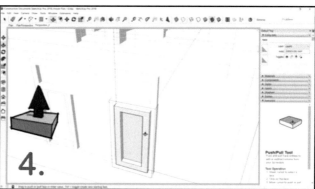

4. Push-pull the frame part outwards by 65mm.

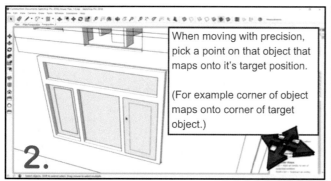

2. Window Leaf moved into place.

When moving with precision, pick a point on that object that maps onto it's target position.

(For example corner of object maps onto corner of target object.)

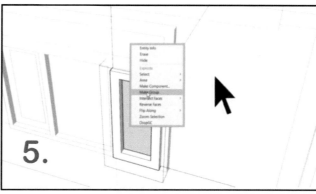

5. Double-click the glass panel surface and select "Make Group" Call the Group "Glass Pane".

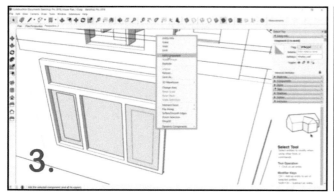

3. Right-click> Edit Component to edit Window Leaf.

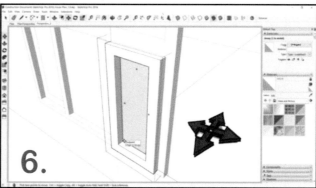

6. Move the glass pane from it's position to the midpoint of the frame

CONSTRUCTION DOCUMENTS USING SKETCHUP PRO 2020

MODELLING THE HOUSE

CREATE THE WINDOW COMPONENT: TEXTURING & GROUPING

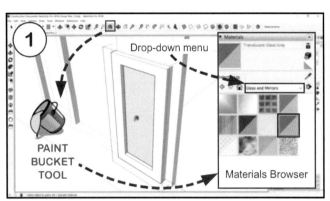

The **Paint Bucket Tool** opens up the **Materials Browser**. Select "**Glass and Mirrors**" from the drop-down menu. Colour the glass pane using grey translucent glass. Close the component when finished.

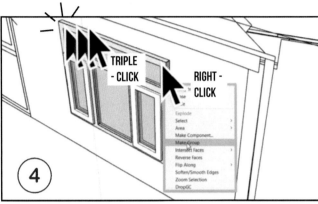

Triple-click on the main window frame to select it.
Right-click and select "Make Group"- We won't name it.

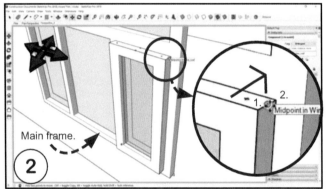

Submerge the small frame back into the main frame: Pick the midpoint of the Window Leaf frame, and position this point at the corresponding corner of the Main frame (2.)

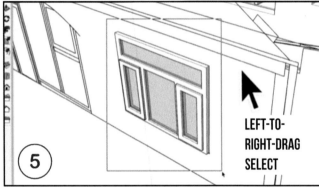

Select all the elements using a left to right drag window selection.

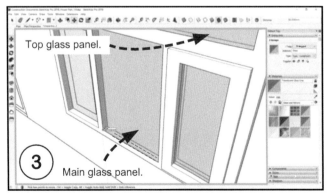

Edit the main glass panel as before and position it in the centre of the Main Frame. Do the same with the top panel. Copy the

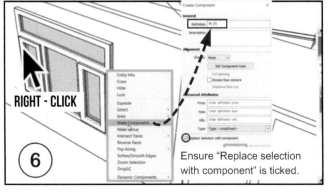

Right-click on the selection.
Click on "**Make Component**". Name it "**W_01**".

CONSTRUCTION DOCUMENTS USING SKETCHUP PRO 2020

MODELLING THE HOUSE

MOVING THE WINDOW COMPONENT INTO PLACE AND FINISHING THE OPE STEP.

A

1. Pick a good "control point" from which to start moving the object.

B

2. Know where you're going to place that "control point".

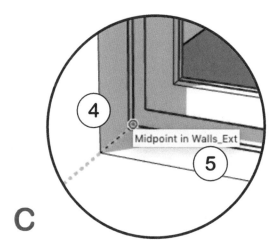

C

Use **Outliner** to examine the structure of your model as it progresses. Try to become familiar with how it works, and how it can help you figure out where you might be going wrong if you're getting stuck.

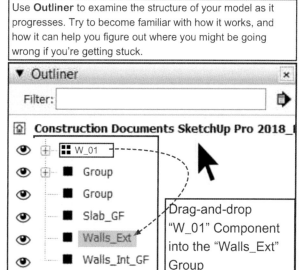

Drag-and-drop "W_01" Component into the "Walls_Ext" Group

To correctly place the window component:

1. Select the object
2. Select the Move Tool.
3. Before moving the object, choose a definite point on it that you know you can position easily.
(Usually a corner point on the front-bottom.)
4. Move the object to the predicted target point.
(Here it's the centre of the wall thickness.)
5. Note the dialogue that displays the location.

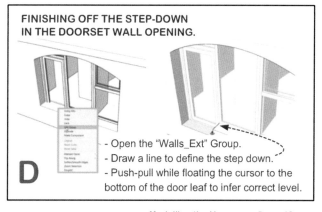

FINISHING OFF THE STEP-DOWN IN THE DOORSET WALL OPENING.

D

- Open the "Walls_Ext" Group.
- Draw a line to define the step down.
- Push-pull while floating the cursor to the bottom of the door leaf to infer correct level.

CONSTRUCTION DOCUMENTS USING SKETCHUP PRO 2020

MODELLING THE HOUSE

INSERTING THE WINDOW AND DOOR COMPONENTS

In this section we're going to populate the model with windows, doors and rear flat roof.

Select the prepared components:

A set of the windows, doors and flat roof have been prepared and ready to insert.

1. Select **Modelling Perspective** Scene.
2. Expand the Components Dialogue
3. Click on the Home button. This displays all the embedded components in the file.
4. Select "_Windows, External Doors & Flat Roof". This component has an "insertion point" that corresponds with the corner point of the extended floor as indicated here.
5. To place, click here.

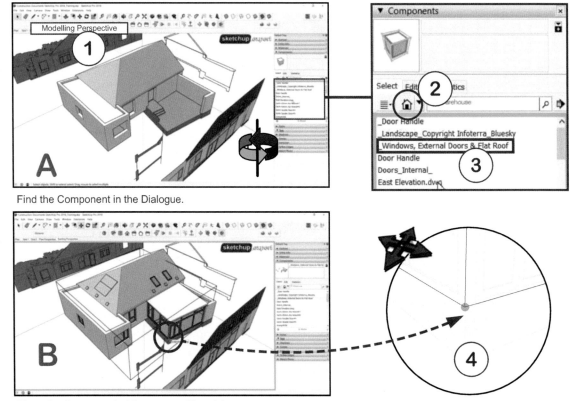

Find the Component in the Dialogue.

Place the Component at the corner of the slab.

CONSTRUCTION DOCUMENTS USING SKETCHUP PRO 2020

MODELLING THE HOUSE

FINISHING OFF THE SLAB

Open the "Slab_GF" component by using Outliner or by double-clicking on it. Tidy up the surfaces and align the edges (if necessary).

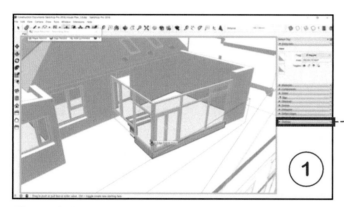

Find the Slab Component in the Dialogue.

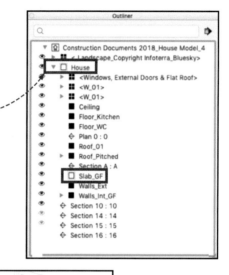

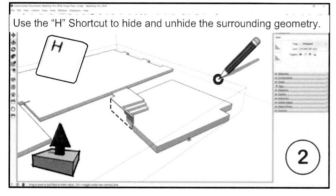

Finish off any gaps using the line tool and push pull the relevant surfaces to align edges.

Finish off the bottom of the slab:

- Delete stray lines.
- Use push-pull to line up surfaces.
- Where holes occur, redraw edges using lines or rectangles.

Neater shapes (shapes that are free of holes or stray lines) are easy to edit and also enable section fill.

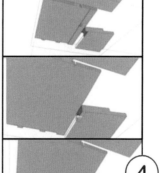

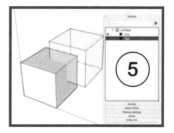

Try out a new feature in Outliner 2020:

Turn off an object (using the eyecon) Click on the hidden object in Outliner.

The object appears in wireframe when selected.
Disappears when de-selected.

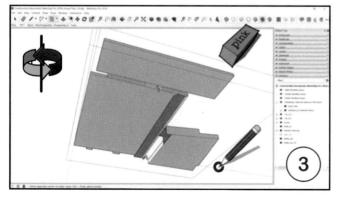

Orbit to the underside of the slab. Finish off missing edges. Delete untidy edges. Finish off surfaces to complete the slab.

CONSTRUCTION DOCUMENTS USING SKETCHUP PRO 2020

MODELLING THE HOUSE

COMPLETING THE INTERNAL WALLS

Next we're going to remove those sections of the internal walls that intersect with the roof. To begin, open the "Walls_Int_GF" group using Outliner.

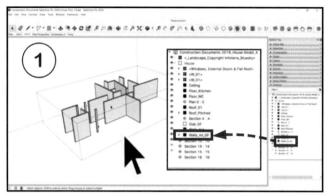

Open **Walls_Int_GF** Group for editing
Select all of the contents inside the Group. (CTRL + A)

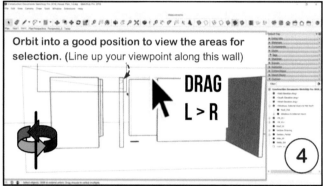

Left-to-Right (drag) selection to select the areas for deletion.
Be careful not to include any areas that you don't want to select.

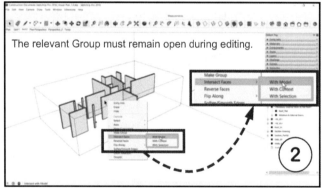

Right-click on selected areas then "**Intersect Faces> With Model.**"

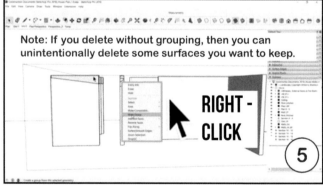

Right-click on the selected area then choose "Make Group"

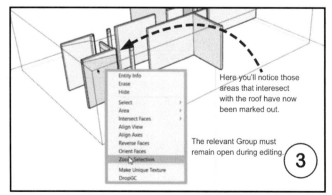

To zoom in on a particular area the "**Zoom Selection**" command is extremely useful.

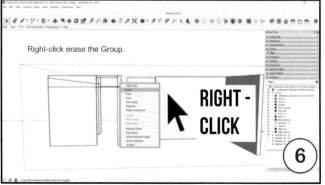

Delete the intersecting volumes. (There are some leftover parts that we'll deal with later.)

CONSTRUCTION DOCUMENTS USING SKETCHUP PRO 2020

MODELLING THE HOUSE

COMPLETING THE INTERNAL WALLS - FILLING IN THE DOOR HEADS

There are two approaches to creating this geometry- Use construction lines to create a framework- or use inferencing.

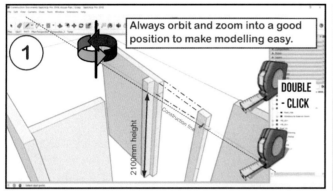

Redraw the edges of the top of the walls (if desired but not necessary.)
Use the tape measure to create construction lines @ 2100mm height.

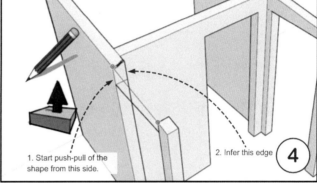

Complete the shape to be extruded

Draw shapes to fill in the volumes. Delete the discarded lines.

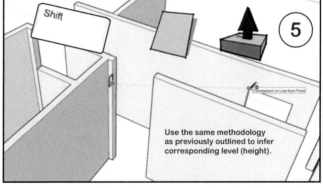

Alternatively use the rectangle tool to draw the extrusion shape.

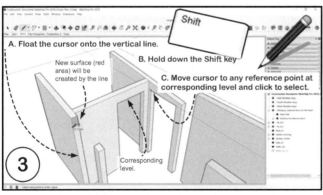

Drawing a line which "splits" a surface to be extruded.
To draw a line at a corresponding level, float the cursor onto a vertical line (1). When the cursor is in place, hold down the Shift button to lock it onto the line (2). As you do this, move the cursor onto the adjacent level and click on it (3).

The blue dots indicate an active surface to be manipulated. Remember- if you don't see these dots appearing it may be because you haven't opened the correct group!

Copyright Paul J. Lee 2020   Modelling the House   Page 53

CONSTRUCTION DOCUMENTS USING SKETCHUP PRO 2020

MODELLING THE HOUSE

COMPLETING THE INTERNAL WALLS - INSERT THE DOORS

To create the internal doors we're going to fetch and insert a component called "**Doors_Internal_**" which is located within the SketchUp file.

Using Outliner lets make sure that we're still editing the **Walls_Int_GF** Group:

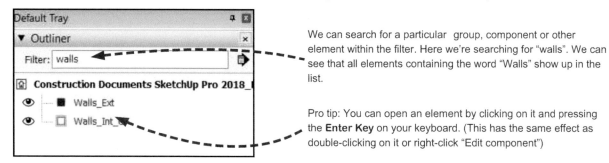

We can search for a particular group, component or other element within the filter. Here we're searching for "walls". We can see that all elements containing the word "Walls" show up in the list.

Pro tip: You can open an element by clicking on it and pressing the **Enter Key** on your keyboard. (This has the same effect as double-clicking on it or right-click "Edit component")

1. Open up the **Components browser**.
2. Click on the home button to view the components that are embedded within the SketchUp file.
3. Use the pull-down menu to view Components as icons or as a list. (Try both views to see the difference.)
4. Find "**Doors_Internal_**" Click on it and then move your cursor into the model space.
5. Place the Component randomly by clicking on any point on the ground plane. (Don't open or edit the Component.)
6. **Pick a strategic point on one of the doors to start moving the component.**
7. Move the component into place.

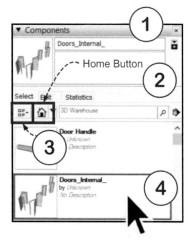

When moving an object always use a near-side corner point on the object. The destination point should be visible without needing to orbit around the model.

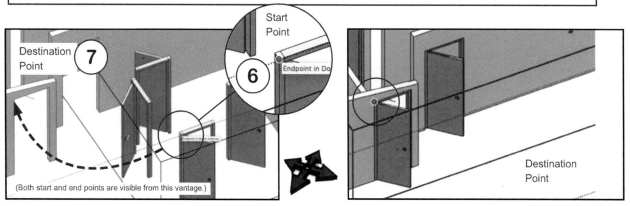

Copyright Paul J. Lee 2020

SETTING UP YOUR DRAWINGS

PLANS SECTIONS & ELEVATIONS: THE SECTION TOOLS

Here we're going to learn about setting up and using the Section Tools.

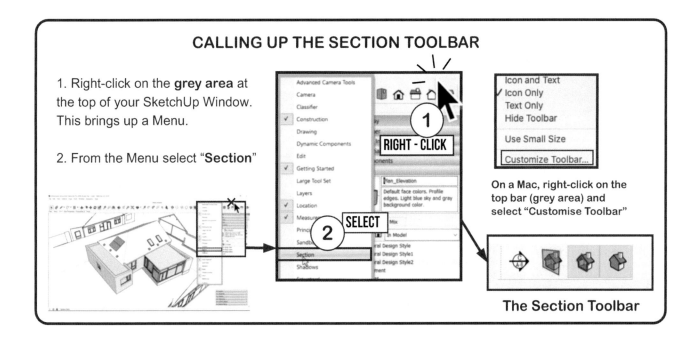

WHAT DO THE FOUR SECTION TOOL BUTTONS DO?:

1. Creates a Section Cut.

2. Makes all Section Frames visible/ invisible.

3. Slices/ Unslices the model.

4. Activates/ Deactivates "Infill".
(This is a new important feature introduced in SketchUp 2018.)

Note: Only an object that is "closed" will allow infill to autogenerate.

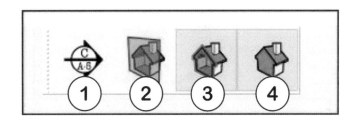

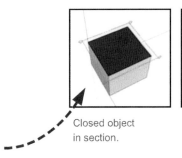
Closed object in section.

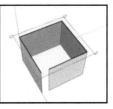
Same object in section but open sided: No infill.

CONSTRUCTION DOCUMENTS USING SKETCHUP PRO 2020

SETTING UP YOUR DRAWINGS

(OVERVIEW) CREATING PLANS SECTIONS & ELEVATIONS

Here's where we start to create our "drawings". To do this we set up "Scenes" (views of the model) in SketchUp, and then we'll import them into LayOut. **This is an overview only**- Detailed steps to follow in the next few pages.

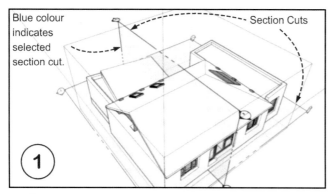

1. The first thing we do is position our "**Section Cut**" These are located and oriented where we want to "slice" our model.

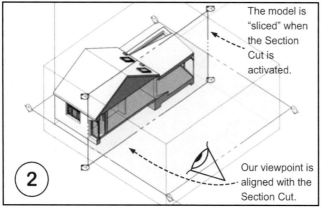

2 Section Cut is activated.

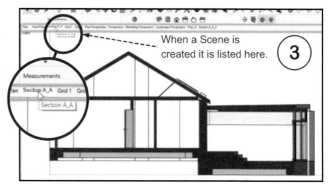

3 **View is aligned** and Scene Created.

4. The Scene is displayed in LayOut- SketchUp's 2D graphic program.
(See the LayOut section of this book for details.)

What is a "Scene"?

"Scenes" are like camera viewpoints from which we view our SketchUp model. They can be set up as Plans, Sections, Elevations, Isometrics, or Perspectives and represented in different **Styles**.

SketchUp Scenes represented in LayOut

CONSTRUCTION DOCUMENTS USING SKETCHUP PRO 2020

SETTING UP YOUR DRAWINGS

PLANS SECTIONS & ELEVATIONS: THE SECTION TOOL

Here we're going to learn about setting up and using the Section Tools.

CREATING SECTION CUTS

TRY THIS ON A BLANK SKETCHUP FILE- (File > New)

Click on the first button to bring up the **Section Cut** mechanism.

- Create a rectangular solid.
- In the Section Toolbar **Select the Section Cut tool** (The first button.)
- Float the cursor around the object and observe how it reacts to the different surface orientations: How it changes colour to Red, Green and Blue.
- Notice how **when you hold Shift it locks in orientation**.
- When you have chosen your orientation, Whilst holding Shift, click on a location on the object.
- Repeat the above process until you have created section cuts for each orientation (Red, Green and Blue)
- Try right-clicking on each Section Cut and selecting "Active Cut"

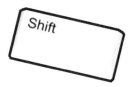

Holding Shift locks in the Section Cut orientation.

You can move selected cuts like any object

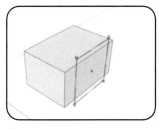 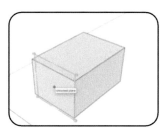  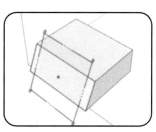

Red, green, blue and magenta (off-axis) orientations.

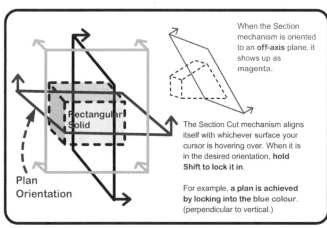

When the Section mechanism is oriented to an **off-axis** plane, it shows up as magenta.

The Section Cut mechanism aligns itself with whichever surface your cursor is hovering over. When it is in the desired orientation, **hold Shift to lock it in**.

For example, **a plan is achieved by locking into the blue colour**. (perpendicular to vertical.)

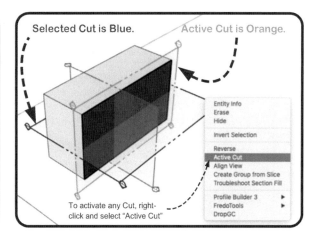

Selected Cut is Blue. Active Cut is Orange.

To activate any Cut, right-click and select "Active Cut"

Copyright Paul J. Lee 2020

CONSTRUCTION DOCUMENTS USING SKETCHUP PRO 2020

SETTING UP YOUR DRAWINGS

PLANS SECTIONS & ELEVATIONS: CREATING THE ORTHOGONAL VIEWS

Here we're going to learn about setting up Scenes that display our Plans Sections and Elevations..

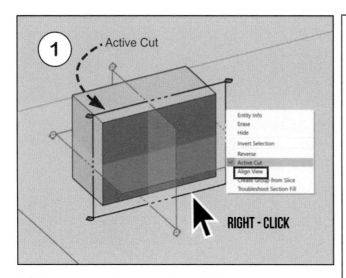

Our Plan, Section and Elevation drawings are orthogonal (Not perspective) drawings. Therefore we must ensure that Parallel Projection is switched on.

To do this, **go to the Camera Menu (top of screen) > Choose Parallel Projection.**

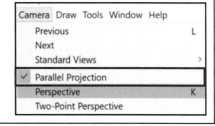

Make sure you have activated the cut you want to set up.
Right-click on the Active Cut and select "**Align View**"

The Aligned View (in Perspective)

The Aligned View in Parallel Projection.)

CONSTRUCTION DOCUMENTS USING SKETCHUP PRO 2020

SETTING UP YOUR DRAWINGS

PLANS SECTIONS & ELEVATIONS: SECTION CUTS & STYLES

Here we'll finish off the setup of the orthogonal view. We want to alter the style of the drawing to our requirements

Click on the second button on the Sections Toolbar to turn off Section Cut visibility

Section Fill is turned on.

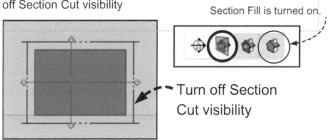

Turn off Section Cut visibility

When the circular arrows appear it means you've made changes that you need to save to the current Style to keep them.

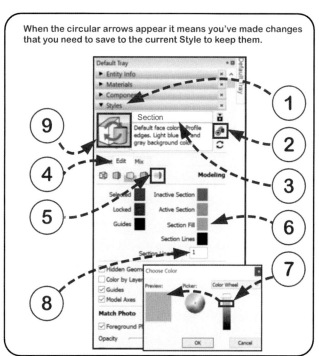

Now we're going to change the appearance of the sliced model in three ways:

- A   Change the infill colour to mid- grey.
- B   Decrease the thickness of the Outline lines.
- C   Change the background to white.

These changes are made in the Styles Console/ Dialogue.

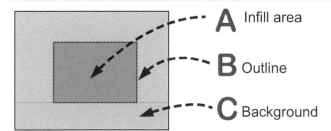

A Infill area
B Outline
C Background

TO CHANGE STYLE (See above right)

1. Open the **Styles Console** in the Default Tray.
2. Click on the "+" button to Create a new Style.
3. Rename the Style to "**Section**"
4. Click on the Edit tab.
5. Where you see the five edit buttons, click on the right-most button.
6. Click on the "Section Fill" colour button.
7. In the Colour Wheel setting, move the slider half way to the top to a **mid grey colour** and click "OK".
8. Change "Section Line Width" to 1 (or leave alone if preferred.)

TO CHANGE THE BACKGROUND COLOUR

- Click on the middle Edit button.
- Click on the Background Colour Square.
- Choose a white colour for the background- Move the slider to the top (See 7 above).
- Click OK.
- Turn off "Sky" and "Ground"

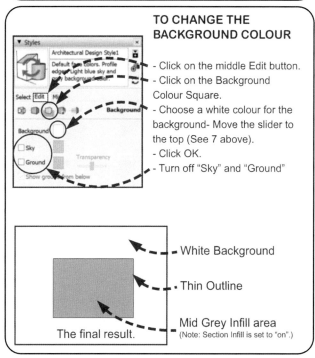

The final result.

White Background
Thin Outline
Mid Grey Infill area
(Note: Section Infill is set to "on".)

Finally, click the update button (9 above)

Copyright Paul J. Lee 2020

SETTING UP YOUR DRAWINGS

PLANS SECTIONS & ELEVATIONS: CREATING SCENES

Scenes are saved views of the model. They also contain **Styles** and multiple other settings. Previously we created our new Style. Now we're going to create a Scene which incorporates that Style.

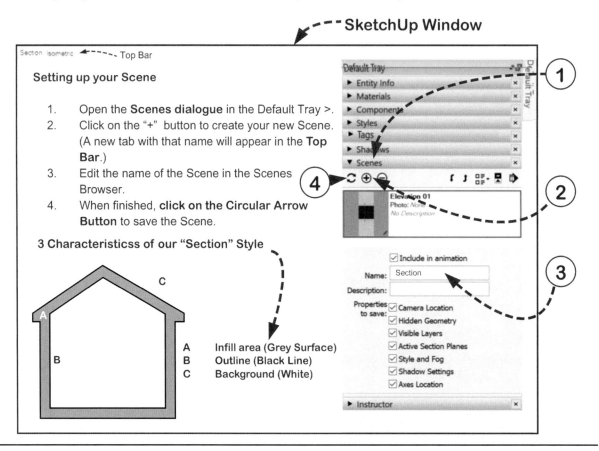

Setting up your Scene

1. Open the **Scenes dialogue** in the Default Tray >.
2. Click on the "+" button to create your new Scene. (A new tab with that name will appear in the **Top Bar**.)
3. Edit the name of the Scene in the Scenes Browser.
4. When finished, **click on the Circular Arrow Button** to save the Scene.

3 Characteristicss of our "Section" Style

A Infill area (Grey Surface)
B Outline (Black Line)
C Background (White)

Updating the Scene is vital before saving the SketchUp file. Otherwise our updates won't appear in LayOut.

When creating a Scene that includes Style changes, a dialogue appears asking whether to create a new Style for that Scene. In general "save as a new style" is the easiest option. However, for our purposes in this case, it is easier if we use one single style called "**Section**" that we'll use in every Plan, Section and Elevation.

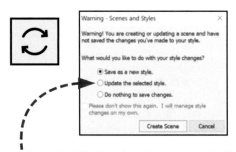

Choosing "Update the selected style" will apply any changes to your current Style and all Scenes where that style is used.

CONSTRUCTION DOCUMENTS USING SKETCHUP PRO 2020

SETTING UP YOUR DRAWINGS

PLANS SECTIONS & ELEVATIONS

It is essential to understand how Section Cuts should be grouped with other elements.

Why?

We want to be able to control (limit) what elements a Section Cut slices through, and what it doesn't slice through. For example, we may want a Section Cut to slice through our building but not slice through the landscape or surrounding buildings.

Below are two diagrams which illustrate the effect of having the Section Cut inside a Group vs. having it outside a Group.

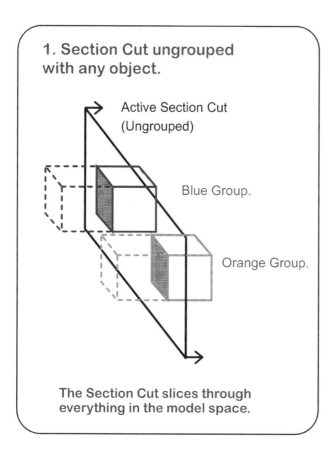

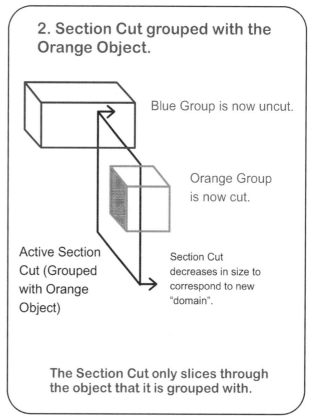

CONSTRUCTION DOCUMENTS USING SKETCHUP PRO 2020

SETTING UP YOUR DRAWINGS

PLANS SECTIONS & ELEVATIONS

Using Outliner here we'll organise our **House Model Elements** into one overall Group so that we can integrate our **Section Cuts** with it.

To begin with, click on the Scene called "**Model Only**" which displays your model with all unnecessary layers switched off. Next, follow the four steps below.

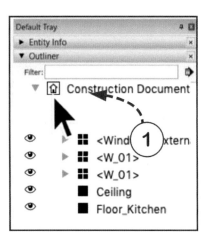

Clicking on the Home Button closes all Groups.

1. Open Outliner. First we're going to **exit all Groups**. The quickest way to do this is to **click on the SketchUp File Name or Home Button**.

3. Next, right-click on the new Group we created and select "Rename". Call the Group "**House**" and hit Enter.

A. CTRL + A*
B. Right-click
C. Make Group.

*(cmd + A for Mac)

2. Next we want to create one Group for our entire House model. To do this in Outliner use **CTRL + A** (to Select All Groups) then right-click on the selection and choose "**Make Group**".
(Note: The Scene displays only the House elements. All other elements are invisible as their layers are switched off.)

4. Next, we want to start placing our Section Cuts within the House Group. Right-click on the House Group and select "**Edit Group**"

CONSTRUCTION DOCUMENTS USING SKETCHUP PRO 2020

SETTING UP YOUR DRAWINGS

PLANS SECTIONS & ELEVATIONS

1. Select the "Model Only" Scene.
2. Call up the **Section Toolbar** as previously outlined.
3. Click on the Section Tool and set up Section Cuts in the positions indicated below. (Plan and Section)

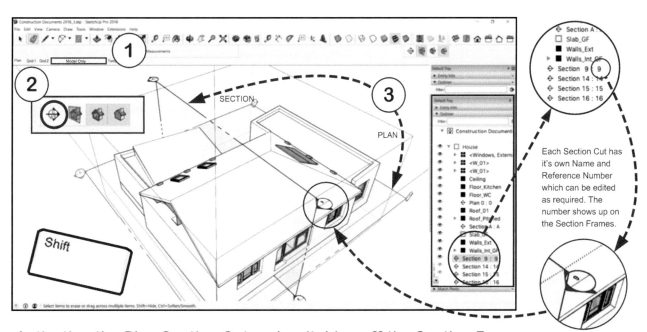

Each Section Cut has it's own Name and Reference Number which can be edited as required. The number shows up on the Section Frames.

Section Entity Number here labelled as "9"

Activating the Plan Section Cut and switching off the Section Frame

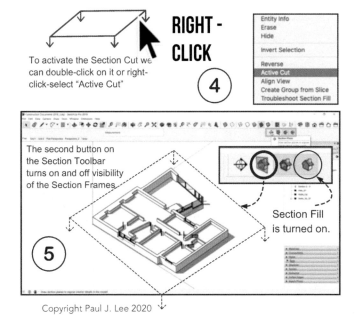

To activate the Section Cut we can double-click on it or right-click-select "Active Cut"

The second button on the Section Toolbar turns on and off visibility of the Section Frames.

Section Fill is turned on.

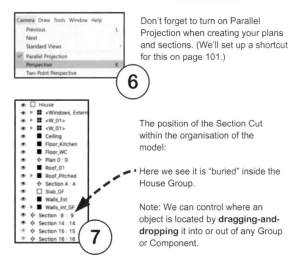

Don't forget to turn on Parallel Projection when creating your plans and sections. (We'll set up a shortcut for this on page 101.)

The position of the Section Cut within the organisation of the model:

Here we see it is "buried" inside the House Group.

Note: We can control where an object is located by **dragging-and-dropping** it into or out of any Group or Component.

Notation for Groups and Components within Outliner are explained earlier in the section called "Structuring Your Model"

SETTING UP YOUR DRAWINGS

PLANS SECTIONS & ELEVATIONS: SCENES

To Activate the Section Cut:

1. Right-Click on it and select **Active Cut**
2. If you need to reverse direction: Right-Click > **Reverse**

To set up the View:

3. Using the **Select Tool** right-click on the Active Section Cut and select "**Align View**"
4. Turn off Section Plane Visibility
5. Ensure that Perspective is Off.
6. Use "**Zoom Extents**" and your mouse wheel to frame the view.

To save the Scene in our SketchUp model setup:

7. Right-click on any one of the Scene Tabs above and select "**Add**"
8. **Rename the Scene as "Section A - A"**
9. **Going down the menu, you can activate the Scenes Dialogue**
10. Using the **Scenes Dialogue** on the right, Clicking on the "**+**" button also creates a Scene.

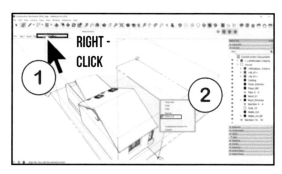

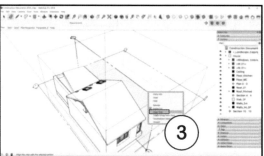

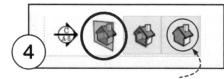

Section Fill is turned on.

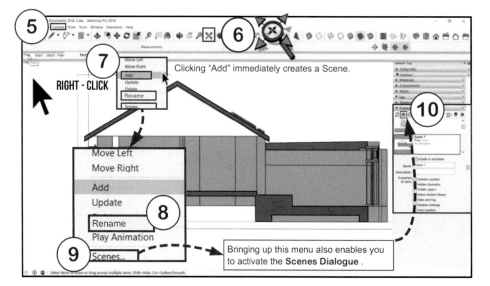

Clicking "Add" immediately creates a Scene.

Bringing up this menu also enables you to activate the **Scenes Dialogue**.

The Scenes Dialogue is thoroughly explained on the following page.

CONSTRUCTION DOCUMENTS USING SKETCHUP PRO 2020

SETTING UP YOUR DRAWINGS

The Scenes Dialogue

This dialogue enables you to create, delete, view and edit your Scenes. It contains many important settings including Camera Location and Styles.

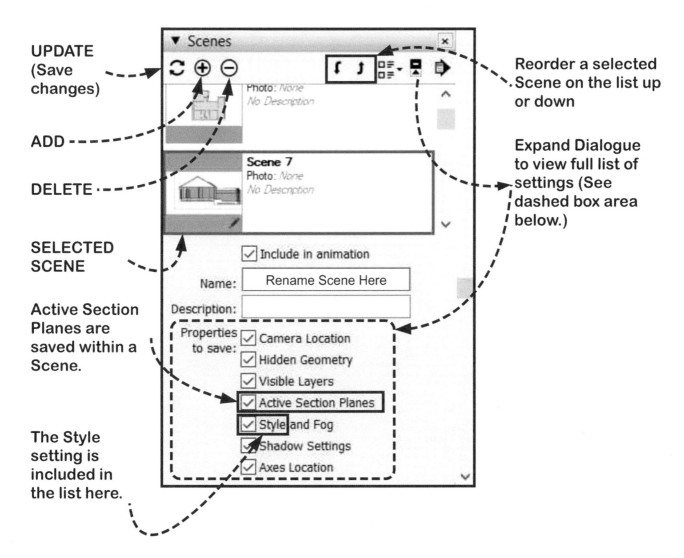

By default, our Scene contains all the settings we have in place at the time of creation. However we can turn each on or off.

SETTING UP YOUR DRAWINGS

PLANS SECTIONS & ELEVATIONS: SCENES

The Plan Scene

1. Using the Section Toolbar, switch on Section Frame visibility (Second Button)
2. Orbit around to give yourself a three quarter view like that shown below
3. Activate the relevant Cut by right-clicking on the Section Frame to select "Active Cut".
4. Right-click on the Section Frame again and select "Align View"
5. Turn off perspective (if required) using the "K" shortcut.
6. Zoom into a good position being careful not to orbit.
7. Turn off the Section Frame Visibility.
8. Create a new Scene by right - clicking on the Scene tabs at the top of the model window and clicking on "Add".
9. You may get a dialogue asking you to Create a new Style. Select "No" to use the "Section" Style.
10. Insert the name "**Plan_New**" in the Name box in the **Scenes Dialogue**.

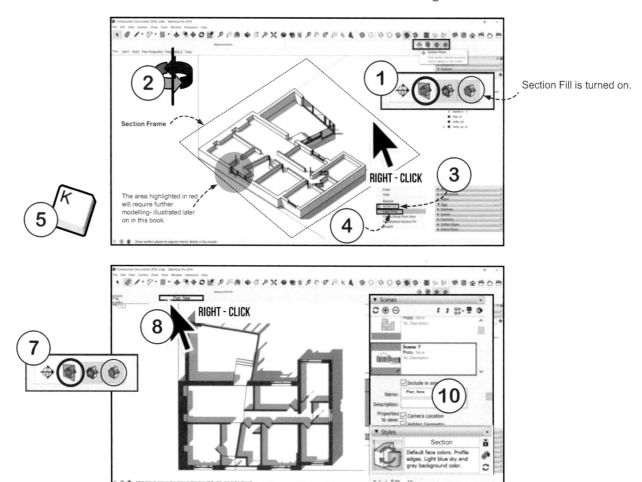

CONSTRUCTION DOCUMENTS USING SKETCHUP PRO 2020

SETTING UP YOUR DRAWINGS

PLANS SECTIONS & ELEVATIONS: SCENES

Setting up Elevations

- **Turn off the Active Section Cut** (if active) to view our model externally.
- **Set up the view** as follows:

We'll use **Standard Views** which are called up in the **Views Toolbar**.

To call up the Views Toolbar: Right-click on the top of the SketchUp Window (Grey Bar) and select "**Views**" from the Menu

Click on the **Front Elevation Button** (See below) to see what we get. (Make sure perspective is still switched off- (Hit "K" to check.)

Create the Scene.

Right-click on any Scene Tab at the top of your screen and select "**Create Scene**".
Right-Click on our new Scene tab and change it's Name to "**West_Elevation**". We'll use this Scene as a template for creating our other elevation Scenes.

In terms of Style, we'll use the "Section" Style (current)

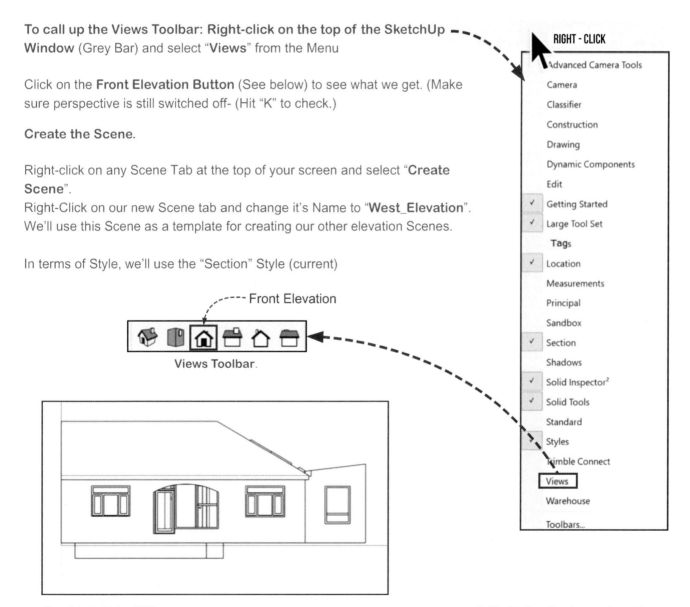

Copyright Paul J. Lee 2020

SETTING UP YOUR DRAWINGS

FINISHING THE MODEL: SETTING UP THE ELEVATIONS

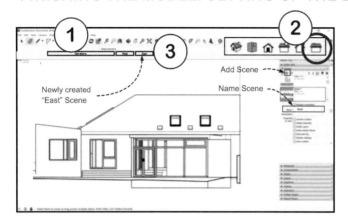

Create the Back (East) Elevation:

1. Use the "Elevations" Scene Tab you created.
2. Using the Views Toolbar, Click on the sixth button from the left to create the Front View.
3. Create a new Scene as described previously and label it as "East_Elevation"

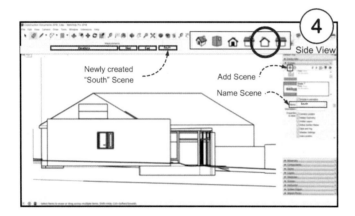

Create the Side (South) Elevation:

4. Using the Views Toolbar, Click on the fifth button from the left (Side View).
5. This time using the Scenes Dialogue, create a new Scene.
6. Name the Scene.

The Scene Update Button (7)

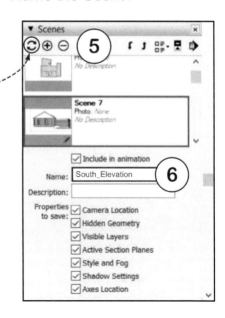

As always, regarding Scenes:

7. If you revise a Scene, always **update it, then** save the SketchUp file again.
8. **Save your SketchUp file after creating or updating a Scene.** (This is vital for updating LayOut drawings as we'll see later.)

SETTING UP YOUR DRAWINGS

TEXTURING THE MODEL

In order to add color and materials to our model, we're going to use the **Materials browser**.

If you don't see "Materials" on the list of menus to the right of your screen in what's called the "Default Tray" then you'll need to add it. This is done by selecting **Window/ Manage Tray/ Default Tray** and then selecting the "**Materials**" tick box in the Dialogs list

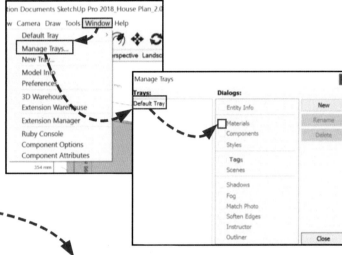

Materials are accessed via the Materials Dialogue or by clicking on the **Paintbucket Tool**.

The Materials Toolbar Elements:

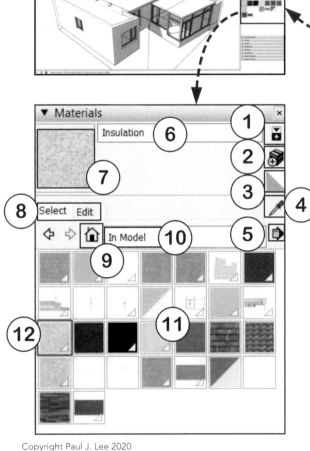

1. Expand Dialogue.
2. Create a New Material Button.
3. Default Material Icon. ("Default Material" is Explained on the next page.)
4. Material Dropper (Copy Material)
5. Select display mode for Current Materials Collection: List by name or show as Icons
6. Current Material Name
7. Current Material Texture/ Colour
8. Select & Edit Tabs.
9. Home Button (Press this to show the Materials Current in your SketchUp Model.)
10. Name of the Current Materials "Collection"
11. Materials in the Current Collection.
12. Current Material is **highlighted** with a blue border.

CONSTRUCTION DOCUMENTS USING SKETCHUP PRO 2020

SETTING UP YOUR DRAWINGS

TEXTURING THE MODEL & ENTITY INFO

There are a number of ways to apply textures.
Materials are applied to:
- Individual surfaces or
- Multiple surfaces at a time.

Below we have three cubes. The first is ungrouped (Raw geometry), and the second two are grouped. Clicking on the Paintbucket Tool allows us to apply a texture (colour) to one face of the **ungrouped** object. If we select another colour and click on the second cube (which is Grouped) the entire object changes to that colour. This is because all faces are set to "Default Colour". Default means that all faces will follow whichever colour is given to a Group / Component.

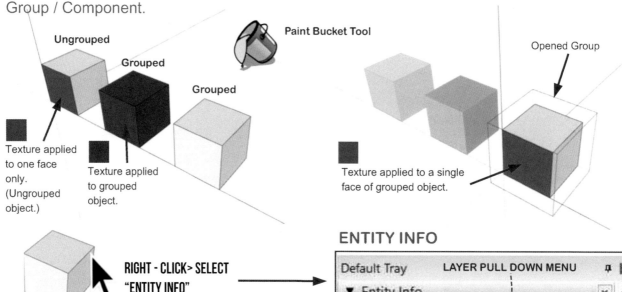

When we select any face, line, or Grouped Object and right-click to call up "Entity Info" we get information on: Material (colour), Layer, and Name of the Object.

If we click on the Colour Icon, we can select a colour for the object. (Note that line colours are controlled in a different way not dealt with here.)

CONSTRUCTION DOCUMENTS USING SKETCHUP PRO 2020

SETTING UP YOUR DRAWINGS

TEXTURING THE MODEL

Now let's texture a few surfaces, starting with the main roof.

1. Call up Outliner and select the "**Roof_01**" Group that we created earlier. Open the Group with a right-click or a double-click.
2. Activate the Paintbucket Tool.
3. In the Materials Browser, select the Home Button
4. Click on the Material called "**Rooftile_Dark**" as illustrated
5. Click on the upper surfaces of the roof to apply the texture.
6. When finished, close the Group. (Remember the quickest way to close all Groups is to click on the main model or "Home" Icon.)

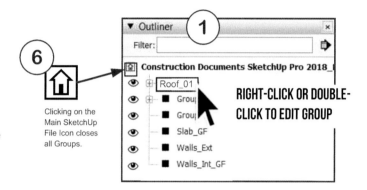

Clicking on the Main SketchUp File Icon closes all Groups.

"ROOFTILE_DARK" MATERIAL

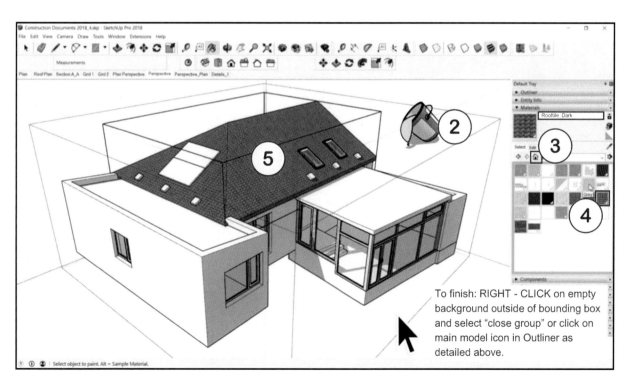

To finish: RIGHT - CLICK on empty background outside of bounding box and select "close group" or click on main model icon in Outliner as detailed above.

CONSTRUCTION DOCUMENTS USING SKETCHUP PRO 2020

SETTING UP YOUR DRAWINGS

TEXTURING THE MODEL

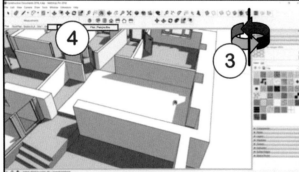
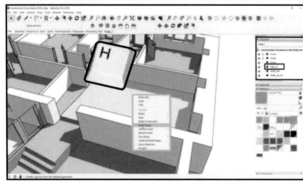
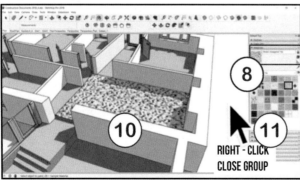

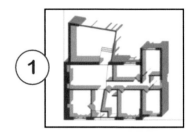

To texture the tiled areas:

1. Open up the view of the interior by clicking on the "**Plan_New**" Scene Tab.
2. Turn on Perspective using "K"
3. Orbit around to (approx) the view shown.
4. Save this as a new Scene. Call it "**Plan_Perspective**".
5. To access the Floor Group, find "**Slab_GF**" in Outliner and open for editing.
6. Draw lines that mark out the areas on the floor surface that you wish to tile such as the kitchen and WCs (Use "H" shortcut to toggle on and off walls to infer.)
7. Activate the Paint Bucket Tool
8. In the Materials Browser select "Tile" from the Collection Pulldown Menu.
9. Select a tile pattern.
10. Click on the selected areas to apply Texture.
11. Close the Group. This time right-click on a blank space outside the Group surrounding box to select "Close Group"
12. **As always, save your model file.**

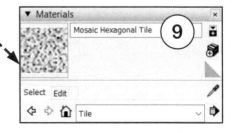

CONSTRUCTION DOCUMENTS USING SKETCHUP PRO 2020

SETTING UP YOUR DRAWINGS

FINALISING THE MODEL FOR LAYOUT

We've set up most of the views that we want, and textured the model. However we haven't included the landscape and surrounding context which we're now going to do. The landscape is already in place on a hidden layer. However, as we can see in the images below when we switch the layer on we can see that it obscures the house. Our Section and Elevation drawings all require that we "slice" through the landscape.

Elevation in isolation

Elevation with landscape

Elevation with "cut" landscape

Landscape layer is switched off.

Landscape layer is switched on.

Landscape layer is switched on.

Adding in the landscape creates the need for extra Section Cuts to cut through the ground plane.

1   Opening the SketchUp file let's go to the Section A - A Scene.
2   Turn on Section Display

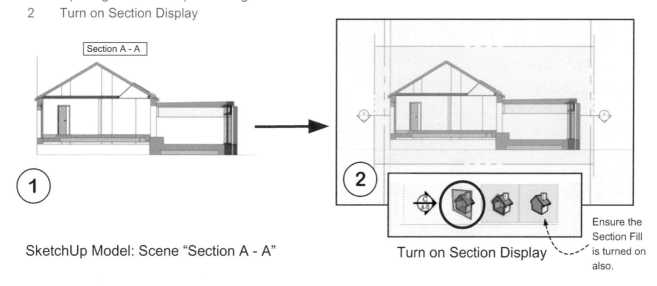

SketchUp Model: Scene "Section A - A"

Turn on Section Display

Ensure the Section Fill is turned on also.

Copyright Paul J. Lee 2020

CONSTRUCTION DOCUMENTS USING SKETCHUP PRO 2020

SETTING UP YOUR DRAWINGS

Notes & Dimensions

In our model space there's an invisible layer which was set up to contain the landscape geometry. We're going to switch on that Tag, create the Section Cut and then line up our view so that it's fully complete for use in LayOut.

1. Before proceeding, ensure that all Groups are closed- Click on the Home Button in Outliner.
2. Orbit and zoom out to (approximately) the viewpoint illustrated here.
3. Expand the Tags Dialogue.
4. Turn on Tag called "A_Hidden_Landscape"

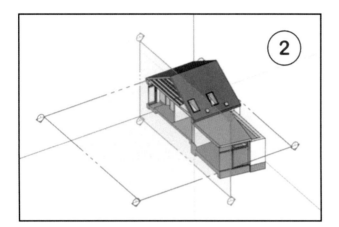

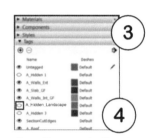

5. Using the House Group for alignment, Create a new Section Cut that will slice through the entire model. (The original Section Cut remains current.)
6. Note the visibility of the new Section Cut in Outliner.
7. Right-Click on the new Section Cut and select "Align View" and then zoom into position.
8. Turn off Section Cut visibility.
9. Right-Click-Update the "Section A - A" Scene.
10. **Save the SketchUp file.**

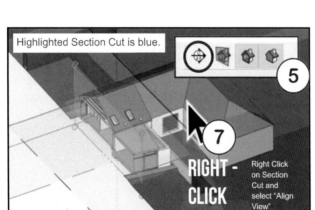

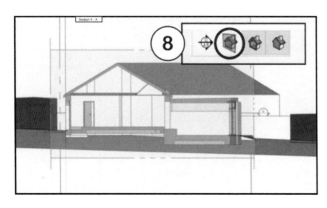

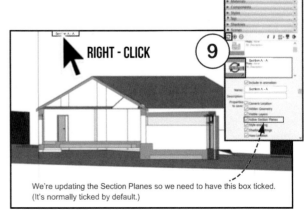

We're updating the Section Planes so we need to have this box ticked. (It's normally ticked by default.)

CONSTRUCTION DOCUMENTS USING SKETCHUP PRO 2020

SETTING UP YOUR DRAWINGS

FINISHING THE MODEL: SETTING UP THE ELEVATIONS

Inserting the landscape into the elevation drawings.

Now that we've set up our three elevations, we need to set up the corresponding landscape in order to finish off the external levels.

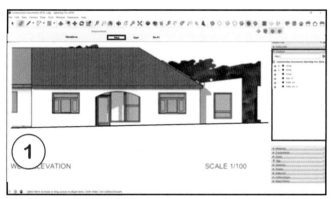

1. Click on the "**West**" Scene that we set up previously.
2. Go to the Layers Browser and click on "**A_Hidden_Landscape**" which will now obscure the view.
3. Bring up the Views Toolbar and click on the Isometric Button.
4. Click on the Section Cut Button and float your cursor onto the front wall of the building and then hold down the Shift Button.
5. Click on the ground plane just in front of the building and you should see the ground sliced away.

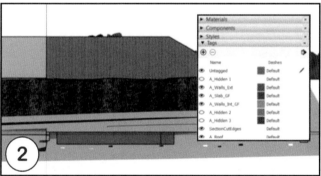
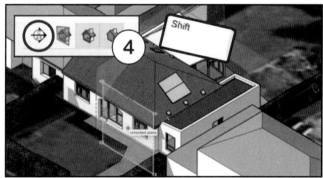

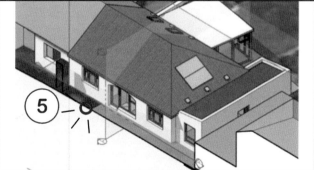

Setting Up Your Drawings

CONSTRUCTION DOCUMENTS USING SKETCHUP PRO 2020

CREATE "POWERFUL SHORTCUT No. 3"

SketchUp saves 5 previous locations from which you were viewing the model. The way to cycle backwards through these locations is via **Camera > Previous**. This is very useful in a number of ways which will become apparent as we proceed.

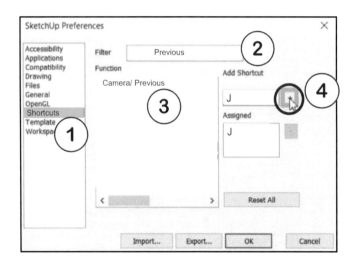

1. Select **Window/ Preferences/ Shortcuts**. (SketchUp > Preferences > Shortcuts for Mac)
2. Type "**Previous**" into the filter search.
3. Click on the selection called: **Camera/ Previous**
4. Type "**J**" into the "Add Shortcut" Dialogue and click on the "**+**" button.
5. If you're asked if it's OK to override an existing function. Select "Yes".

Now the shortcut is ready. Test the effectiveness of the shortcut by hitting the the "J" key and see how the 5 previous views are displayed in sequence.

CONSTRUCTION DOCUMENTS USING SKETCHUP PRO 2020

SETTING UP YOUR DRAWINGS

FINISHING THE MODEL: SETTING UP THE ELEVATIONS

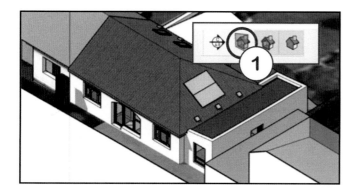

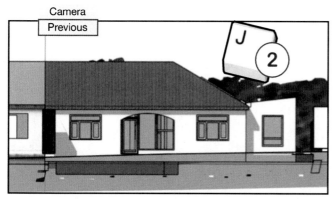

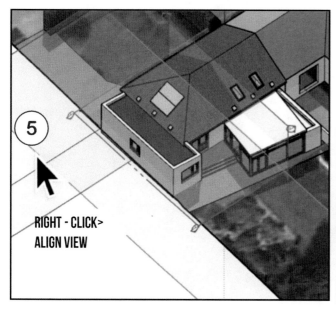

Having set up the Section Cut for our Elevation:

1. Click the Section Cut Display Button to turn off the Cuts.
2. To return to the view back to where we had it previously, hit the **"J"** key. Keep pressing until you get to the Elevation view you had previously.
3. To save the changes to the Scene, click on the Update Button on the Console.
4. Repeat the process for the other two elevations.
5. An alternative approach to setting up our elevation we can use the Section Cut as a reference: Right-Click on the Section Cut and select "**Align View**".
6. Remember to resave the Scene.

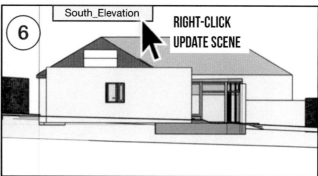

CONSTRUCTION DOCUMENTS USING SKETCHUP PRO 2020

SETTING UP YOUR DRAWINGS

FINISHING THE MODEL: SETTING UP THE PERSPECTIVES

To finalise the Scenes we'll use a Scene Tab that has no Section Cuts: The Landscape Perspective.

1. Click on the Landscape Perspective Scene Tab. (This Scene is saved with all Section Cuts switched off and all layers switched on.)
2. Orbit to your preferred viewpoint.
3. Right-click on the Scene Tab and select "Add Scene"
4. Right-click on the new Tab and select "Rename"
5. Call this Scene "Perspective 1"
6. Repeat to create further Perspectives 2, 3 etc..
7. As always, save the SketchUp file.

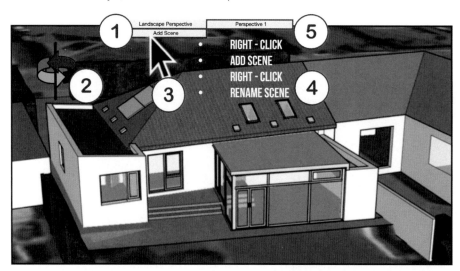

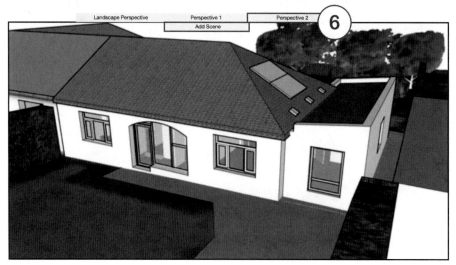

CONSTRUCTION DOCUMENTS USING SKETCHUP PRO 2020

MODELLING THE HOUSE: FINISHUP

There were a couple of things we needed to finish in the model. They were:

- Remodelling to the entrance area of the house
- Completion of the party (North) wall.
- Creating the Ceiling Group
- Fixing the floor slab (if needed).
- Create openings for Skylights in the roof.

THE ENTRANCE AREA: WALL

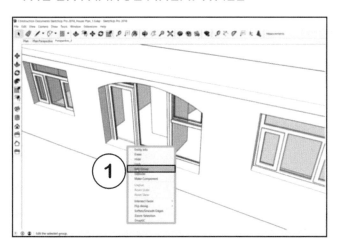

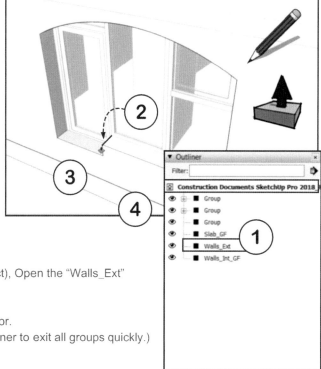

To model the step down in the wall at the front entrance:

1   Using Outliner (or using your mouse to right-click-select), Open the "Walls_Ext" Group.
2   Draw a line onto the surface of the Wall as indicated.
3   Push-pull the surface to level with the bottom of the door.
4   Close the Group (Hint: Click on the House Icon in Outliner to exit all groups quickly.)

THE ENTRANCE AREA: FLOOR

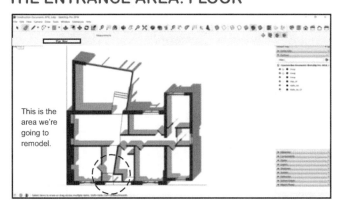

This is the area we're going to remodel.

To model the step down at the floor in the front entrance:

We're going to get a clear view of the floor by clicking on the **"Plan_New"** Scene that we created previously.
Turn on Perspective: Hit "K" (Shortcut we set up earlier)

Go to the next page.

CONSTRUCTION DOCUMENTS USING SKETCHUP PRO 2020

MODELLING THE HOUSE: FINISHUP

THE ENTRANCE AREA: MODELLING THE FLOOR SLAB

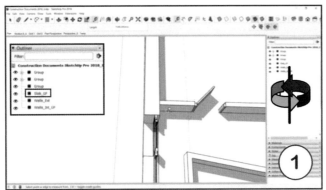

1. Orbit around to the front entrance. Open Outliner and edit the Ground floor slab **"Slab_GF"**

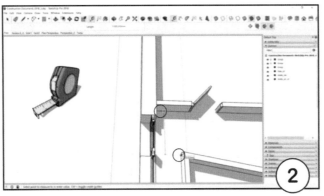

2. Create a construction line that coincides with the width of the front door.

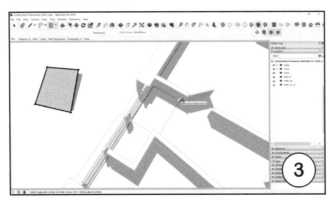

3. Draw a rectangle to reflect the shape we've just outlined.

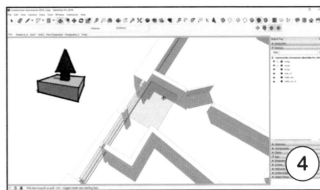

4. Use push pull tool to create a hole in the slab.

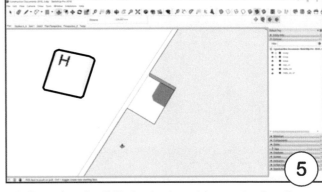

5. Pressing the "H" shortcut enables us to view the slab in isolation.

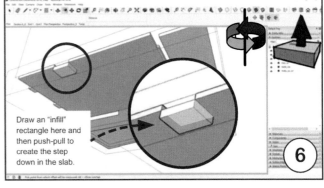

6. Orbit to the underside of the slab. Draw a new rectangle into the location as shown here and create a step down.

MODELLING THE HOUSE: FINISHUP

THE ENTRANCE AREA: MODELLING THE FLOOR SLAB & THE GABLE WALL

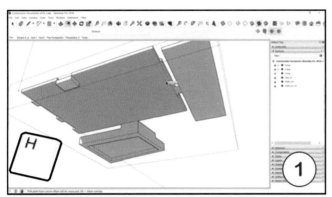

1. Offset the outlines of the slab inwards by 300mm

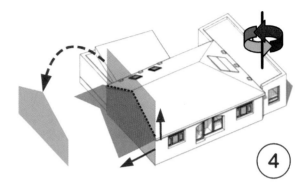

4. Now we're going to shape the gable wall by extending it up and using the roof to "slice" it.

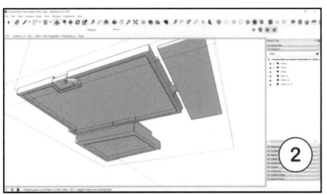

2. Push-pull the "border" area to line up with the bottom of the step.

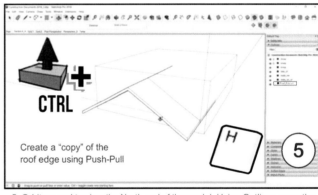

5. Orbit around to view the North end of the model. Using Outliner, open the "**Roof_01**" group. Holding down the CTRL key, push-pull a surface from the roof edge and close the Group.

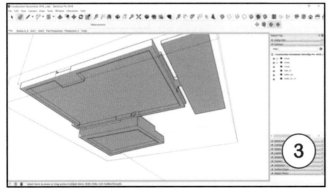

3. Delete all unnecessary lines.
When finished, **exit all Groups** by clicking on the Home Button in Outliner
(Next we're going to edit the Gable Wall.)

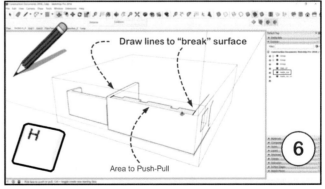

6. Open the "**Walls_Ext**" Group from Outliner. Draw lines on the top of the gable wall as indicated.

CONSTRUCTION DOCUMENTS USING SKETCHUP PRO 2020

MODELLING THE HOUSE: FINISHUP

THE ENTRANCE AREA: MODELLING THE FLOOR SLAB & THE GABLE WALL

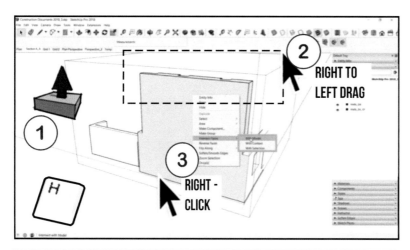

1. Push-pull the top wall surface up above the level of the roof surface (The roof isn't visible here as we're using the "hide" function.)
2. Select the surfaces that you need to slice (Use a right-to-left drag rectangle selection)
3. Right-click on the surface and select "Intersect With Model"
4. The result is the "splitting" of the wall surfaces.

To delete the surfaces we don't need:

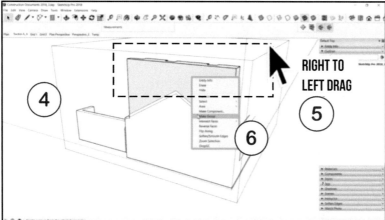

5. Using the Select Tool, drag from right to left across the surfaces to highlight them as indicated.
6. To make them easy to delete we can make them a group (This means we don't accidentally delete lines that other surfaces depend on.) To do this, Right-click on the selection and pick "Create Group"
7. With the Group highlighted, click on the **Delete** button (On your keyboard.)
8. With all surfaces deleted, Close the Group
9. Delete any stray lines or surfaces.

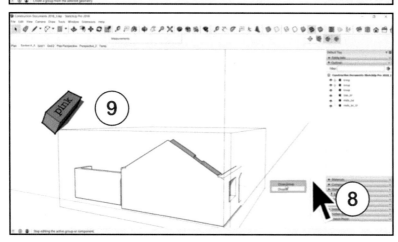

CONSTRUCTION DOCUMENTS USING SKETCHUP PRO 2020

CREATE "POWERFUL SHORTCUT No. 4"

SketchUp has a special paste function that is specific to 3D modelling context. It allows us to copy an object and paste it exactly into position. We're now going to create a shortcut for this using the letter "Z".

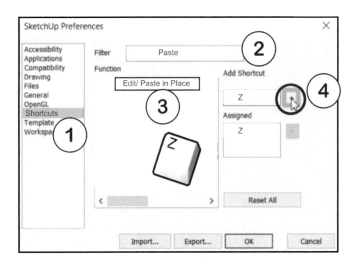

1. Select **Window/ Preferences/ Shortcuts**. (SketchUp > Preferences > Shortcuts for Mac)
2. Type "**Paste**" into the filter search.
3. Click on the selection called: "**Edit/ Paste In Place**"
4. Type **"Z"** into the "Add Shortcut" Dialogue and click on the **"+"** button.
5. If you're asked if it's OK to override an existing function. Select "Yes".

Now our shortcut is ready. Test it's effectiveness:

1   Start a new SketchUp file.
2   Create a box.
3   Select and copy one of it's surfaces (CTRL + C)
4   Start another SketchUp file.
5   Press "Z"
6   The surface should be in the exact same place in the new file.
7   Test it's positioning by copying another of the cube's surfaces in the same way repeating the process. The surfaces should line up perfectly.

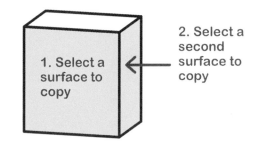

This shortcut should be used constantly throughout your day-to-day modelling. Note that it can also be used between different files. This is a very useful technique for modelling a large SketchUp file to speed up modelling: Take a piece of geometry you're working on and paste it into a new file. Keep working on it there and then copy paste in place back into the original file.

CONSTRUCTION DOCUMENTS USING SKETCHUP PRO 2020

CREATE "POWERFUL SHORTCUT No. 5"

SketchUp generally works best when viewing your model in perspective. We're going to set you up with a shortcut that toggles between orthographic and perspective.

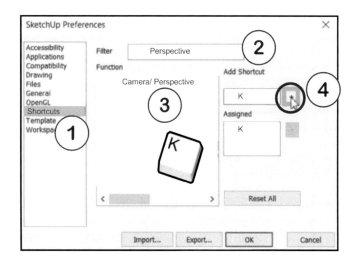

1. Select **Window/ Preferences/ Shortcuts**. (SketchUp > Preferences > Shortcuts for Mac)
2. Type "**Perspective**" into the filter search.
3. Click on the selection called: "**Camera/ Perspective**"
4. Type "**K**" into the "Add Shortcut" Dialogue and click on the "**+**" button.
5. If you're asked if it's OK to override an existing function. Select "Yes".

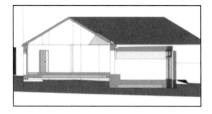

Click on "Section A - A" or Plan/ Elevations.

Now our shortcut is ready. Test it's effectiveness:

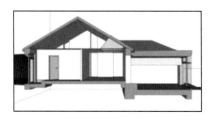

Hit "K".

Click on any orthographic scene in your model (Plan/ Section/ Elevation)
Hit the "K" key.
Notice how the drawing becomes a perspective.
Hit the "K" key again.
You're back to Parallel View.

Hit "K" again.

This shortcut will be used constantly throughout your day-to-day modelling.

CONSTRUCTION DOCUMENTS USING SKETCHUP PRO 2020

MODELLING THE HOUSE: FINISHUP

MODELLING THE CEILING

To create the ceiling, we're going to copy the floor surfaces which are the same shape.

1. Click on the "Plan_New" Scene
2. Using Outliner open "Slab_GF" Group. (Or double-click twice to drill down into the Group.)
3. Select the floor surfaces. Using Select Tool, while holding down Shift to select multiple surfaces, select all floor surfaces. Don't select the Sun Room extension floor surface.
4. Use CTRL + C to copy.
5. Close the Slab Group. (But don't close the House Group. The ceiling needs to be kept inside the Group.)
6. To generate the next vantage point Click on the "Section A - A" Scene.
7. Press "Z" (Our Paste In Place Shortcut)
8. Turn on Perspective: Use the "K" shortcut to bring up Perspective View.
9. Create a Group (The pasted geometry is still selected) Right Click > Make Group
10. Using Outliner or Right Click Entity Info Name the Group as "Ceiling" (Check that it's inside the House Group.)
11. Move the ceiling upwards into position as shown. **Use the Up Arrow on your keyboard to lock movement into the blue (vertical) axis.**
12. Open the Ceiling Group.
13. Press "H" shortcut to view the ceiling slab in isolation.
14. Turn off Section Cut to see the full Group.
15. Delete all internal dividing lines (if any)
16. Using Push-Pull Give it a thickness of say 25mm for a double-slab plasterboard. Close the Ceiling Group.

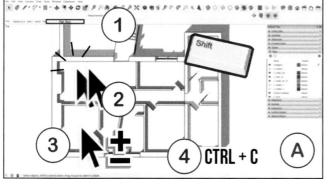

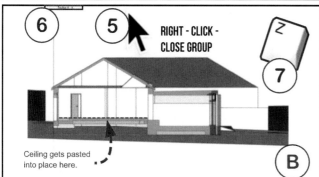

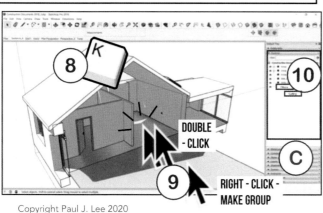

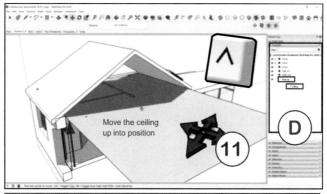

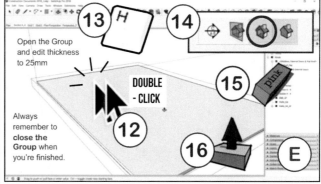

CONSTRUCTION DOCUMENTS USING SKETCHUP PRO 2020

MODELLING THE HOUSE: FINISHUP

MODELLING THE ROOF OPENINGS AND "SKY FUNNELS"

We're going to cut holes in the roof surfaces that correspond with the roof lights.
To begin, select the "Model Only" Scene and navigate to a view approximating (1) below.

> Open the "Roof_01" Group using Outliner.
> Select the Rectangle Tool.
> Holding down Shift Button lock the cursor onto the roof surface
> Use the roof light as an inference to draw a rectangle from corner to corner onto the roof surface.
> Orbit around to the far side to place the second point.

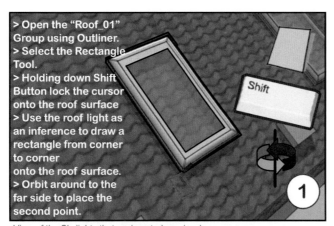

View of the Skylights that we inserted previously.

To create an opening:

Using the Select Tool: Double-click on the shape.

Right-click and Make it a Component

(Previously for doing this we just used the Push-Pull Tool. Here we have "tricky geometry" that isn't doing what we want so we're using a different method to create the opening.)

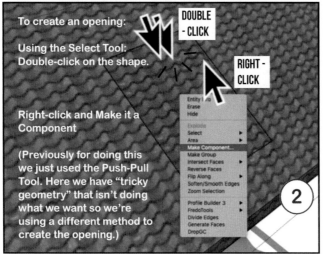

Name this Component "Top Roof Opening".

Ensure that "Cut Opening" is ticked. "Replace selection with Component" should be ticked.

Create Component.

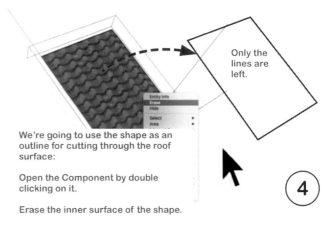

We're going to use the shape as an outline for cutting through the roof surface:

Open the Component by double clicking on it.

Erase the inner surface of the shape.

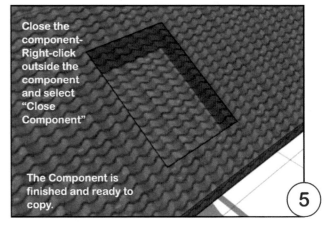

Close the component- Right-click outside the component and select "Close Component"

The Component is finished and ready to copy.

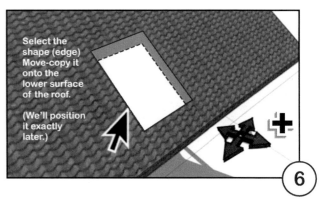

Select the shape (edge) Move-copy it onto the lower surface of the roof.

(We'll position it exactly later.)

CONSTRUCTION DOCUMENTS USING SKETCHUP PRO 2020

MODELLING THE HOUSE: FINISHUP

MODELLING THE ROOF OPENINGS AND SKY "FUNNELS"

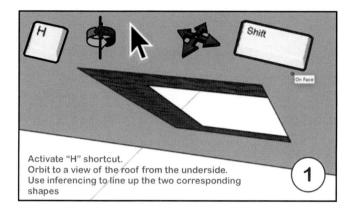

1. Activate "H" shortcut.
Orbit to a view of the roof from the underside.
Use inferencing to line up the two corresponding shapes

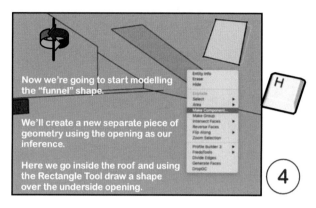

4. Now we're going to start modelling the "funnel" shape.

We'll create a new separate piece of geometry using the opening as our inference.

Here we go inside the roof and using the Rectangle Tool draw a shape over the underside opening.

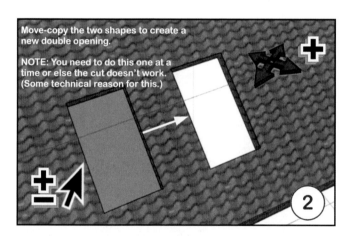

2. Move-copy the two shapes to create a new double opening.

NOTE: You need to do this one at a time or else the cut doesn't work. (Some technical reason for this.)

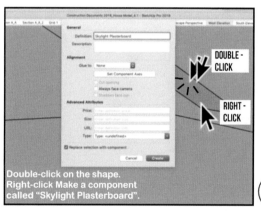

5. Double-click on the shape.
Right-click Make a component called "Skylight Plasterboard".

3. Pressing "H" shortcut makes the skylights reappear.
This allows for lining up the openings.

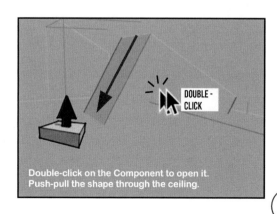

6. Double-click on the Component to open it.
Push-pull the shape through the ceiling.

CONSTRUCTION DOCUMENTS USING SKETCHUP PRO 2020

MODELLING THE HOUSE: FINISHUP

MODELLING THE ROOF OPENINGS AND SKY "FUNNELS"

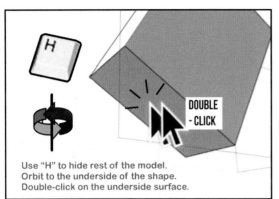

(1) Use "H" to hide rest of the model.
Orbit to the underside of the shape.
Double-click on the underside surface.

Drag from top-right to bottom left.

(4) With a Right to Left Crossing Selection, select the area as indicated above.
Right-click on the selection and select Make Group.
Delete the Group. Now the funnel shape is complete.

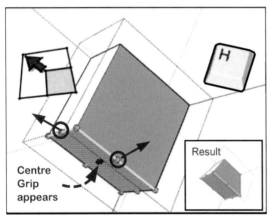

(2) Click on the Scale tool to make the Grips appear. (Green boxes.)
Holding down the CTRL button (Alt for Mac) Pull across the centre grip to expand the shape size.
Do this in both width and length directions. Use "H" to turn on and off the building view.

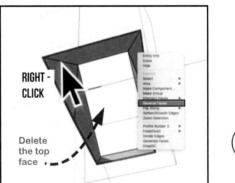

(5) Now we're going to switch the orientation of the surfaces (They're inside-out. We want the inside faces to appear white.)
Select the entirety of the shape as before.
Right-click and select "Reverse Faces".

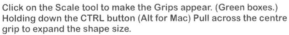

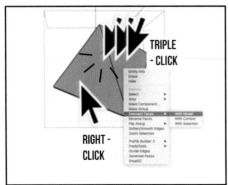

(3) Hide environment using the "H" button.
Triple-click on the shape to Select All - or - Use CTRL + A - or - Use a L - R capture selection. Right-click on the selected shape and select "Intersect faces with Model"

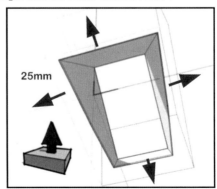

(6) Faces are orientated as desired.
Push-pull the surfaces outwards by 25mm to create a thickness for each side.
Close the component.

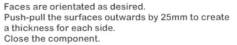

CONSTRUCTION DOCUMENTS USING SKETCHUP PRO 2020

MODELLING THE HOUSE: FINISHUP

MODELLING THE ROOF OPENINGS AND SKY "FUNNELS"

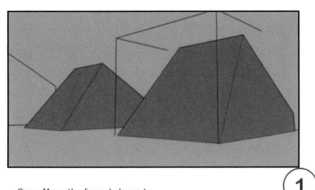

Copy-Move the funnel shape to line up with the adjacent opening. (1)

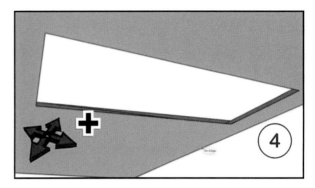

Copy the opening to the upper surface of the ceiling. The opening is complete. (4)

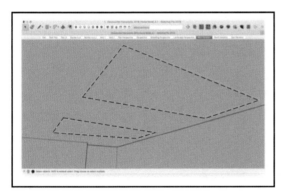

Orbit to the underside of the ceiling. You can see the outline of the funnels cutting through the ceiling above. Open the Ceiling Group.
Draw the shape of the outline of one of the funnels onto the ceiling. (2)

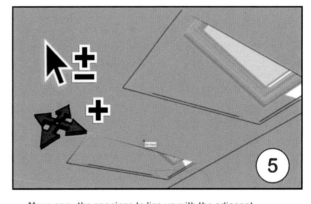

Move-copy the openings to line up with the adjacent funnel.
Close the Ceiling Group. (5)

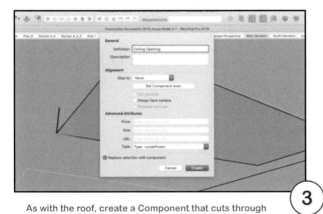

As with the roof, create a Component that cuts through the ceiling surface.
Delete the internal surface. (3)

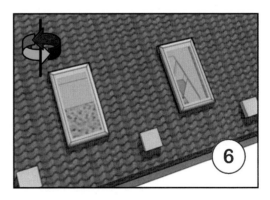

Orbiting to to look at the top of the roof we can see right through to the floor inside. (6)

CONSTRUCTION DOCUMENTS USING SKETCHUP PRO 2020

MODELLING THE HOUSE: FINISHUP

Let's take a look at our model:

1. Close all Groups.
2. Click on the "Landscape Perspective" Scene.
3. Save your model (CTRL + S)

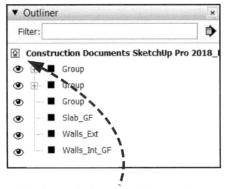

To close all Groups, Click on the "Home" button in Outliner.

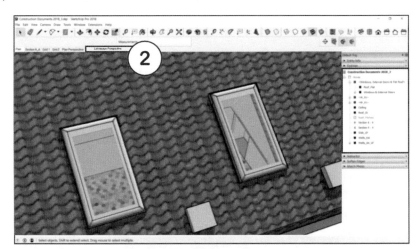

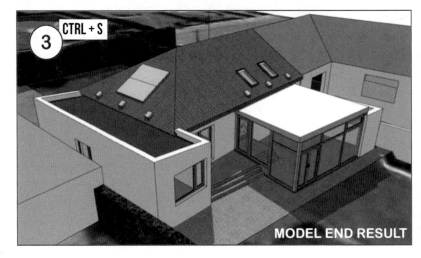

MODEL END RESULT

If you've reached this point successfully congratulations! We've now completed the overall building model and are ready to tackle the detail models and finally the drawings in LayOut.

If you've reached this point successfully congratulations! We've now completed the overall building model and are ready to tackle the detail models and finally the drawings in LayOut.

CONSTRUCTION DOCUMENTS USING SKETCHUP PRO 2020

MODELLING THE DETAILS

DETAILS MODEL FILE

In your downloaded files you'll find the file called "**Construction Documentation SketchUp Pro_Details_X.X**"

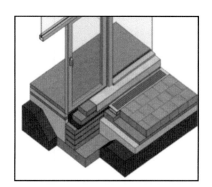

With this file we're going to build the details that accompany our House Model.

Here's what our Model file looks like:

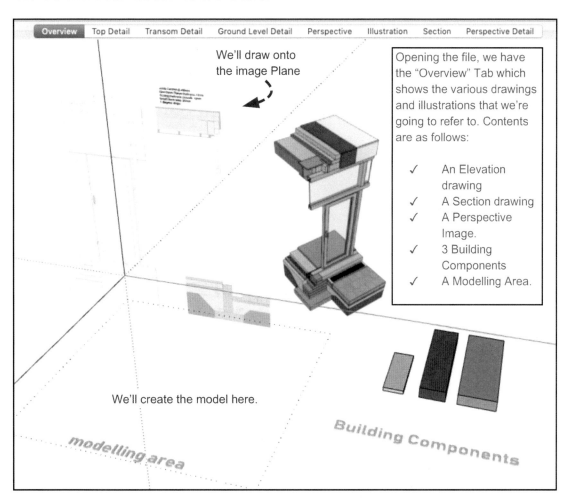

We'll draw onto the image Plane

Opening the file, we have the "Overview" Tab which shows the various drawings and illustrations that we're going to refer to. Contents are as follows:

- ✓ An Elevation drawing
- ✓ A Section drawing
- ✓ A Perspective Image.
- ✓ 3 Building Components
- ✓ A Modelling Area.

We'll create the model here.

MODELLING THE DETAILS

OBTAINING A SECTION DRAWING FROM THE MODEL

Let's go back to our House Model.

To create our drawing, we're going to generate a wireframe from the model as follows:

1. Click on the Scene "Modelling Perspective"
2. Switch on Section Display
3. Using Outliner, Right-Click on Section A - A
4. Select "Create Group From Slice"
5. This Group appears in Outliner.
6. Copy it (CTRL + C)

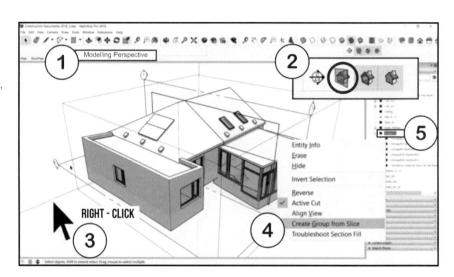

Now lets go to the Details Model.

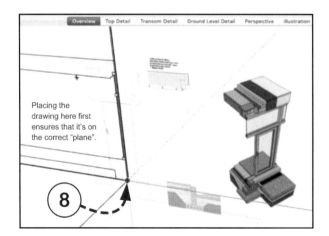

7. Paste in the Section Group that we copied from the Model.
8. Place it on the origin as illustrated here.
9. Move it into place over the Section Drawing.

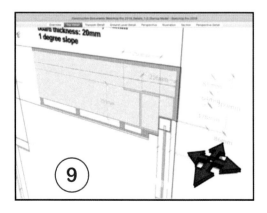

Float the line drawing over the image as illustrated. We'll use these lines as the basis to form geometry for building the details.

CONSTRUCTION DOCUMENTS USING SKETCHUP PRO 2020

MODELLING THE DETAILS

CREATING THE EXTRUSIONS FROM THE DRAWING

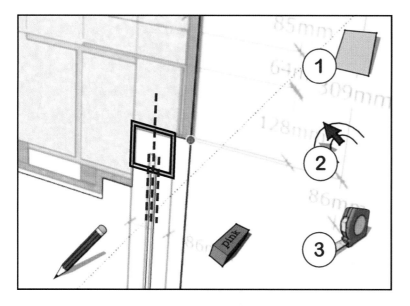

Let's draw the top part of the glazing (Transom)

1. Using the Rectangle Tool, trace over the drawing to create a square.
2. Use Offset to create the inner square @3mm thickness.
3. Use the Tape Measure Tool to find the mid-point of the shape to set out the cut for the glazing (6mm)

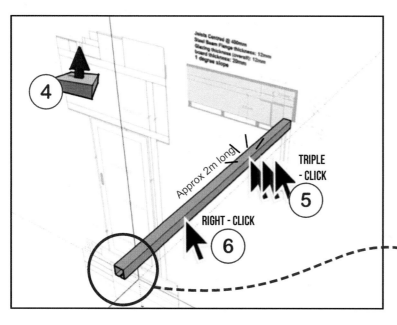

4. Extrude the part to about 2m length.
5. Using Select Tool triple-click to select the geometry.
6. Right-Click and make it a Group.

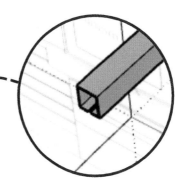

Copyright Paul J. Lee 2020

CONSTRUCTION DOCUMENTS USING SKETCHUP PRO 2020

MODELLING THE DETAILS

BUILDING ELEMENTS FROM THE DRAWING

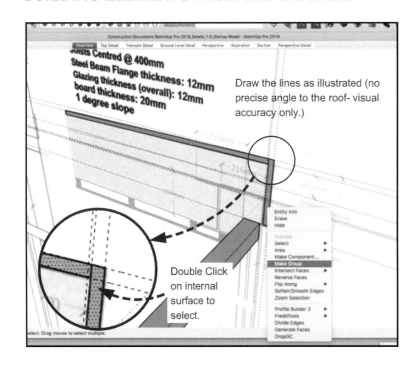

1. Using the Line Tool trace over the boards to create separate regions as illustrated.
2. Double click the internal region of a shape to select it.
3. Right-Click Make Group with each one.
4. Open Groups to extrude (See illustration below.)
5. Close Groups.

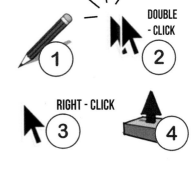

6. Using the Paint Bucket Tool activate the Materials Browser.
7. Click on the Home Button to view the native list of materials.
8. Choose Metal and Wood for the elements as shown.

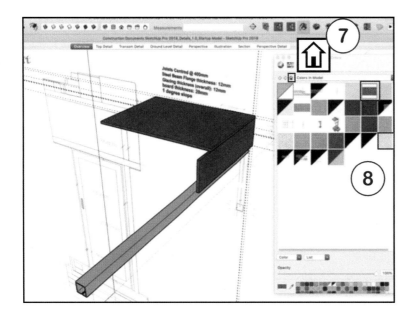

CONSTRUCTION DOCUMENTS USING SKETCHUP PRO 2020

MODELLING THE DETAILS

BUILDING ELEMENTS FROM THE DRAWING

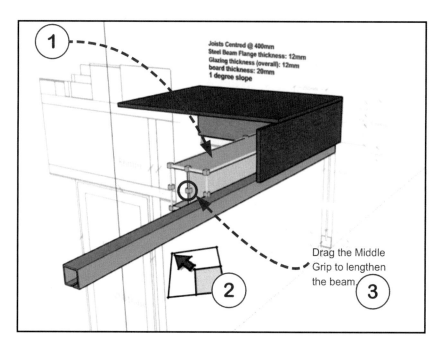

1. Activate the Components Browser on the right of your screen and import the "250mm H-Beam" into position. (Click on the Home button to bring up native Components.)
2. With the beam selected, activate the Scale Tool.
3. Drag the middle handle outwards to lengthen the beam to the same length as the mullion.

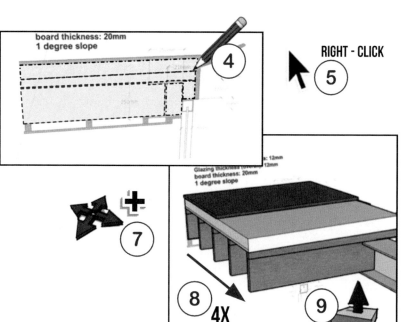

4. Draw the shapes of the rafters and "firring pieces" that fit around the H-Beam.
5. Create separate Groups for the Rafter and the Firring Piece.
6. Selecting both Groups create a Component, called "50 x 250mm Rafter".
7. Use Move-Copy to create a copy of the Rafter @ 400mm as specified.
8. Type "4X" and hit Enter to create an array.
9. Draw and extrude the insulation.

Copyright Paul J. Lee 2020   Modelling Details   Page **95**

CONSTRUCTION DOCUMENTS USING SKETCHUP PRO 2020

MODELLING THE DETAILS

MODEL THE INSULATION, BATTENS & CEILING

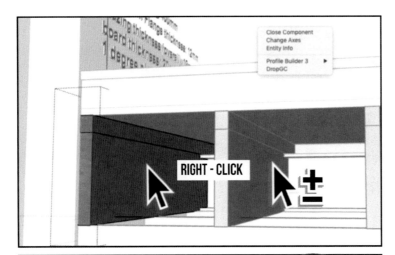

To Create the Insulation Battens:

1. Taking the rafters we created, let's re-open the Rafter Component.
2. Select both Groups inside, holding down Shift to do this.
3. CTRL + C to copy them.
4. Close the Component.
5. Paste in Place ("Z" shortcut.)
6. Move the elements into place.

7. **Texture both pieces:**
8. Using the Paint Bucket Tool while **holding down alt** to activate Sampler, take a sample of the insulation batten that's located in the "Building Components" section.
9. Paint the texture onto both Groups.
10. As they're selected, incorporate both Groups into a new component called "Insulation Battens"
11. Move-Copy the new Component into place as illustrated.
12. Type "3x" and hit Enter to create an array of the Battens.

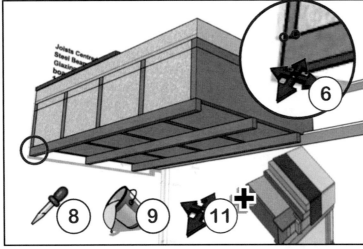

13. **To create the underside Ceiling Battens,** draw a 50 x 40mm rectangle in place.
14. Extrude the rectangle to the length indicated.
15. Triple-click/ Right-click to create a Component called "Ceiling Batten".
16. Sample and use the "Wood" Texture as described above.
17. Move-Copy to array the battens as indicated@ 435mm centres.

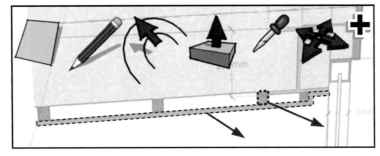

18. **Draw the Ceiling profile and** extrude it.
19. Texture using the Plaster texture this time from the Materials Browser.

CONSTRUCTION DOCUMENTS USING SKETCHUP PRO 2020

MODELLING THE DETAILS

IMPORT AND POSITION VARIOUS COMPONENTS

We're going to use a few in-built Components to create our details. Their names are listed in the illustration below. Using the Components Browser, click on the Home button to find the list of Components. Import the 5 listed below. Place them casually first and then use the drawing to finally position them.

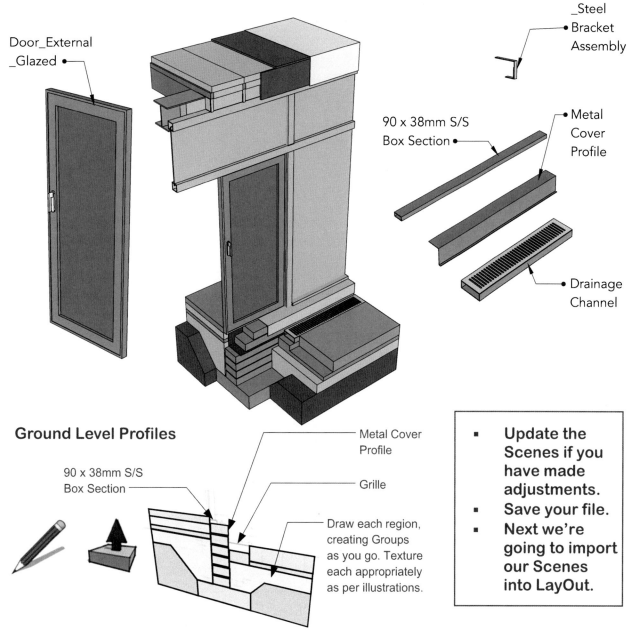

- **Update the Scenes if you have made adjustments.**
- **Save your file.**
- **Next we're going to import our Scenes into LayOut.**

CONSTRUCTION DOCUMENTS USING SKETCHUP PRO 2020

DRAWING PRODUCTION: LAYOUT

LayOut is the graphic interface for SketchUp models.

LayOut's vector line capabilities include scaled drawing functionality which facilitate export to PDF, DWG, DXF as well as .jpg and .png file types. Menus and dialogues contain a vast array of options for customising lines, linetypes, shapes, infills, dimensions and text.

SketchUp models are scaled within LayOut. Dimensions, graphics and notes are then applied. This section of the book illustrates using LayOut to achieve final drawing production from our model.

Here is the first of the sheets we're going to produce:

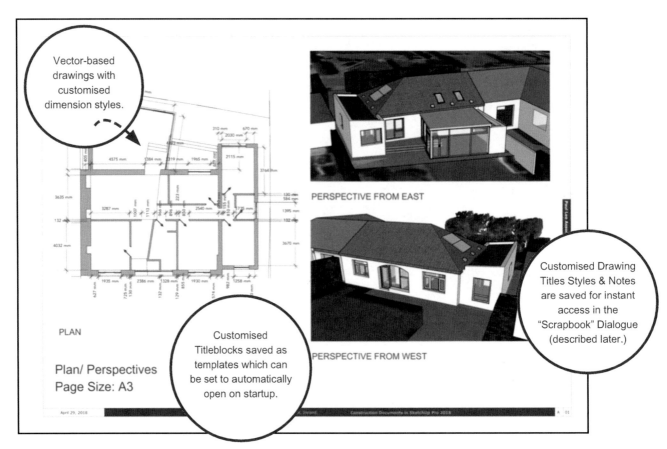

DRAWING PRODUCTION: LAYOUT

On starting up, LayOut initiates the dialogue below.
We have two main choices:

1. Open a pre-set template (There are various categories to choose from.) or….
2. Select from the previously saved files listed.

Let's select Titleblock- A3 Landscape

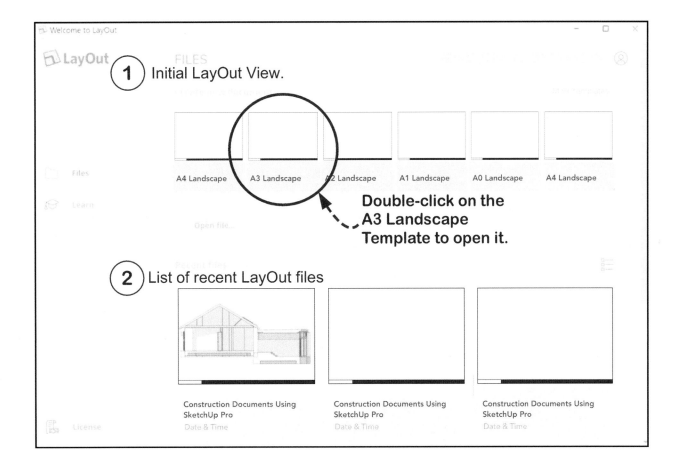

CONSTRUCTION DOCUMENTS USING SKETCHUP PRO 2020

DRAWING PRODUCTION: LAYOUT

INSERTING THE SKETCHUP MODEL

To insert a Model:

1. **Go to: File > Insert.**
2. Select our SketchUp file that we've been working on.
3. The SketchUp Window appears in LayOut showing the view of the model that was current when we saved it.
4. Right-click on the window and go to **Scenes** to select the Scene you wish to view from the pull down menu.

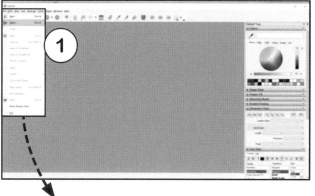

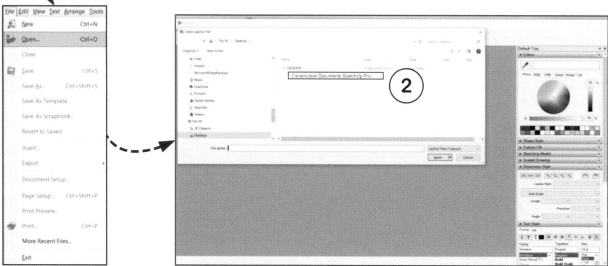

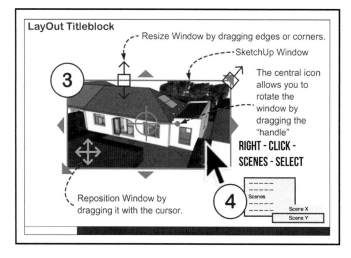

Resizing the Window:

With the SketchUp Window selected (highlighted), float the mouse around the Window and notice the different icons appearing over the model and on the edges. Drag the edges or corners to resize the window as illustrated in the example here.

Repositioning the Window

When you float the mouse over Model Window the Crosshairs appears. When this happens, hold the mouse button down to drag the Window into position.

CONSTRUCTION DOCUMENTS USING SKETCHUP PRO 2020

DRAWING PRODUCTION: LAYOUT

CALLING UP THE "A - A" SCENE:

To present a Scene that we've prepared in SketchUp:

1   Right- Click on the SketchUp Model Window
2   Select "Section A - A" from the dropdown menu. The Scene should now be displayed.

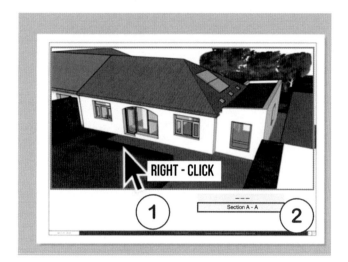 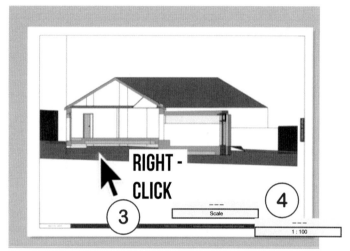

Selecting a Scale for the Scene.

3   **Right-Click on the Section A - A Scene** as above and select "**Scale**" from the dropdown menu.
4   Select "1/100". (Note that when a scale is chosen, we can then resize the window without the scale being changed. This setting can be turned off to allow dynamic rescaling- We'll see how that works later.)

LayOut allows us to display our models to scale and to apply accurate graphics.

We're going to scale our Model Scenes and then apply graphics, dimensions and notes. To do this we need to accurately **infer** linework in the drawings. For this we require **Object Snap to be switched on.** It's normally switched on by default but lets check:

Go to **Arrange > Object Snap** to see that it's ticked.

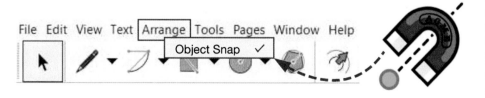

Object Snap is required to "Grip" lines and endpoints in your SketchUp Model.
It enables us to accurately apply graphics and dimensions to our model.

LAYOUT: DIMENSIONS

TO CREATE DIMENSIONS: Click on the **Dimension Tool** button (1) and then click on your two reference points (2, 3). The next point (4) will locate the text and the arrows. Select the dimension to activate move function (5).

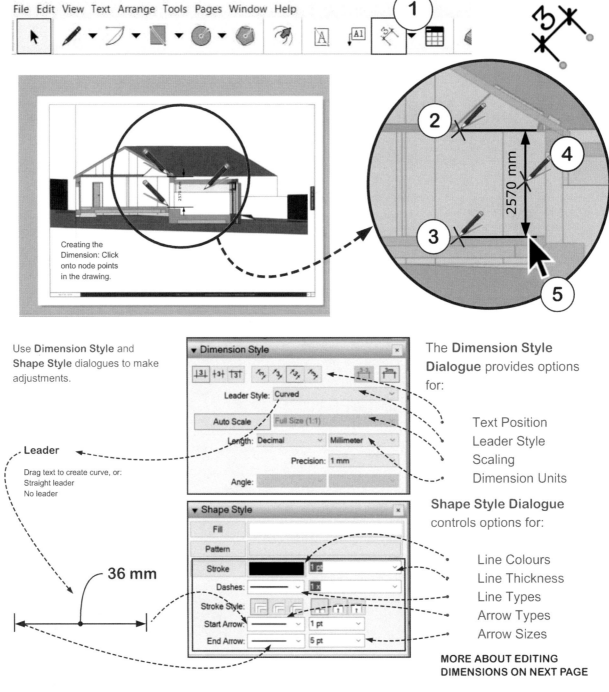

Use **Dimension Style** and **Shape Style** dialogues to make adjustments.

Leader

Drag text to create curve, or:
Straight leader
No leader

The **Dimension Style Dialogue** provides options for:

- Text Position
- Leader Style
- Scaling
- Dimension Units

Shape Style Dialogue controls options for:

- Line Colours
- Line Thickness
- Line Types
- Arrow Types
- Arrow Sizes

MORE ABOUT EDITING DIMENSIONS ON NEXT PAGE

CONSTRUCTION DOCUMENTS USING SKETCHUP PRO 2020

LAYOUT: DIMENSIONS

Orientation and positioning of text.

On selecting a dimension object, it will display various grips (blue colour) that allow it to be repositioned.

Drag the blue (highlighted) elements to reposition the Dimension.

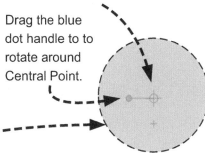

Central Point (Can be dragged into new position as an inference point.)

Drag the blue dot handle to to rotate around Central Point.

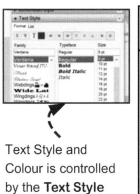

Text Style and Colour is controlled by the **Text Style Dialogue**.

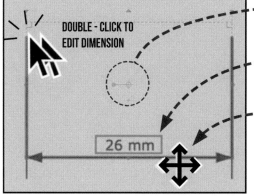

Selected (highlighted) dimension

Drag text to reposition and auto-generate Leader

Float mouse over dimension to see cross appear. Drag dimension to reposition.

Dimensions will automatically update in response to changes in the associated SketchUp model.

Updates in LayOut 2019

Angled Dimension Leaders

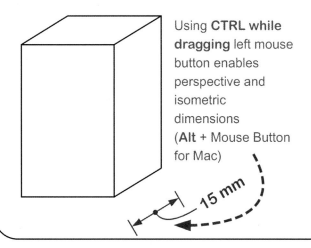

Using **CTRL while dragging** left mouse button enables perspective and isometric dimensions (**Alt** + Mouse Button for Mac)

Bespoke Text with Auto-Dimensions

1. Double-click into text.
2. Write your text **before** the "<>" symbol.
3. Click away.

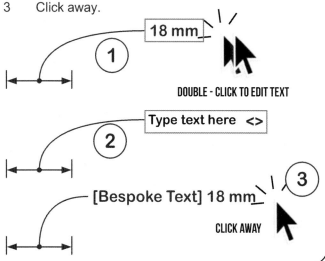

CONSTRUCTION DOCUMENTS USING SKETCHUP PRO 2020

LAYOUT: SHAPES & INFILL

Here we're going to create a shape using the SketchUp model's node points to generate a shape for the ground that is cut away.

1. Click on the **Line Tool** and then click on all the **node points** in your drawing that relate to the cut in the ground.
2. Complete the profile shape as illustrated.
3. Open the Shape Style Dialogue on the right.

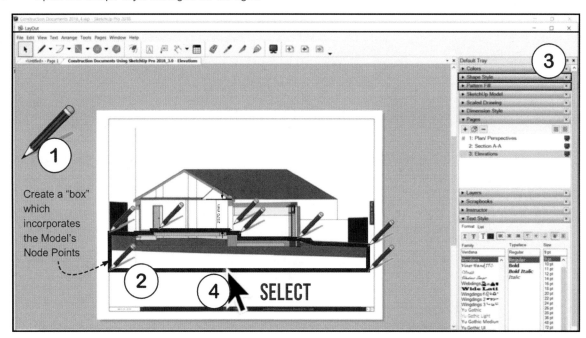

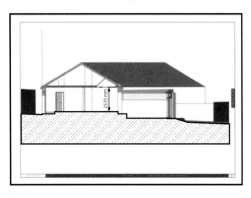

4. Select the outline shape for editing.
5. Turn on Fill.
6. Stroke Colour: Turn on and Select Black.
7. Stroke Thickness: Set to 1 Pt.
8. Click on the **Pattern Button**.
9. Select "Material Symbols" from the dropdown menu.
10. Select "Brick Common Face"
11. Select Scale: 0.25

CONSTRUCTION DOCUMENTS USING SKETCHUP PRO 2020

LAYOUT: TEXT
SETTING OUT THE TEXT

To start creating text click on the text button or hit "T" shortcut.

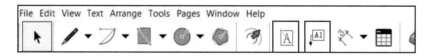
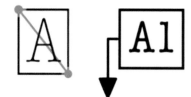

Creating text is generally done in one of three ways:

1. Basic line input
2. Setting out text boxes.
3. Using Call-outs

1. Single-click and type text. Click away to finish.

2. Create a text box by dragging and then type. Click away to finish.

3.
- Click on the call-out icon
- Select a start point, mid point and text input point. (A, B, C)
- Click away to finish.

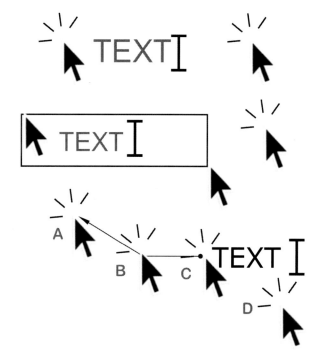

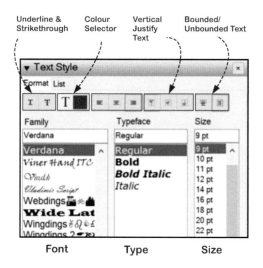

The Text Style Dialogue presents options for colour, font, size and others. **This interface controls text used in Dimensions also.**

Text Style can be edited by single-clicking on it and making changes via the Dialogue.

Text content is edited by double-clicking on the text box, editing and then clicking away to finish.

CONSTRUCTION DOCUMENTS USING SKETCHUP PRO 2020

LAYOUT: POSITIONING & SIZING
SETTING OUT THE TEXT

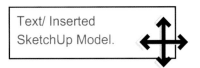

Moving the Text:
As you float your mouse over the text, the cursor changes to a cross. Reposition the text box by dragging on it.

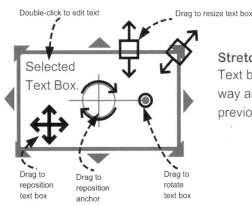

Stretching & Rotating Text:
Text boundaries are reshaped, moved and rotated in the same way as shapes. They can also have fills and boundaries (See previous section regarding "Shapes".)

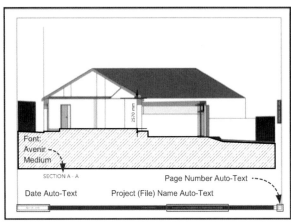

<Auto-Text>

Auto-Text is used to automatically change in response to inputs such as page number, page name, date etc. It is written as a kind of "code" like this:

<text inside brackets>

Practical examples are
<PageNumber>   Auto-displays page number.
<PageName>   Auto-displays page name.
<DateCreated>   Auto-displays date created.

Note: The text must be written exactly with first letter as a capital with no spaces

You can edit auto text inputs by clicking on the Top Page Menu:

Go to: **Text > Customise Auto-Text**

There you'll see the listed **auto-text options** in the Document Properties Console.

You can also create your own Auto Text categories by clicking on the "+" button. There are also "duplicate" and "-" buttons.

Auto-Text Console.

Copyright Paul J. Lee 2020   LayOut: Text 2   Page **106**

CONSTRUCTION DOCUMENTS USING SKETCHUP PRO 2020

DRAWING PRODUCTION: LAYOUT
SETTING OUT AN A3 SHEET

Below are some of the typical font and dimension sizes for an A3 size sheet.

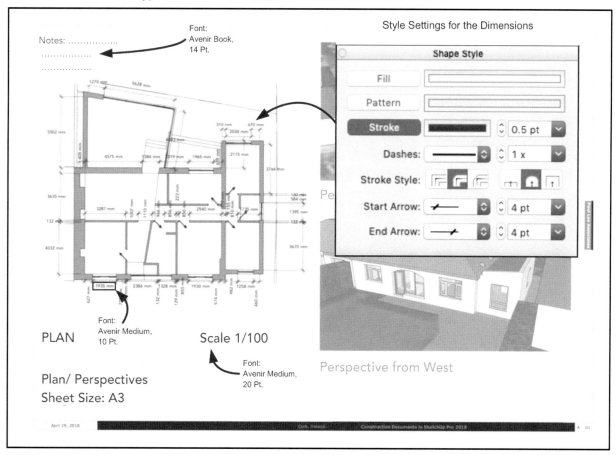

COPYING A SKETCHUP WINDOW

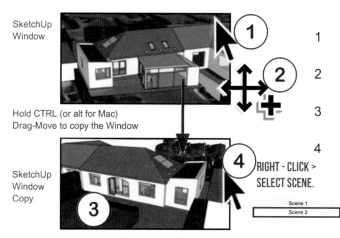

1. Float your cursor over a window until you see a crosshares.
2. Hold down CTRL key (You should see a "+" sign next to the cursor.) (Alt for Mac)
3. Drag the copied Window into position and release the mouse first.
4. To change the Scene, just right-click to show the drop-down menu to select a new Scene.

Copyright Paul J. Lee 2020

DRAWING PRODUCTION: LAYOUT

SKETCHUP MODEL

LayOut's Dialogues provide many powerful features. There's a new look for how the SketchUp Model is displayed. Clicking on the model window and expanding the SU Model Dialogue you'll get the following options to select from:

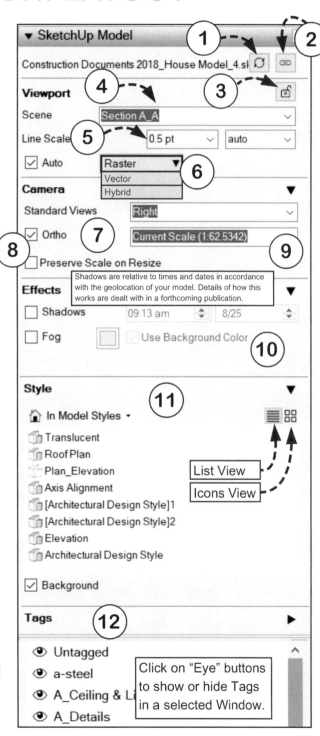

1. New: Refresh button (Click on this to display changes in your updated model.)
2. New: Click to browse for a new model reference for your selected window.
3. New: You can now lock your model window so that it doesn't get moved or edited.
4. Select Scene from those within the SU environment you created. (Pull-down menu.)
5. Your preferred line weight.
6. Pull-down menu: **Render Modes: Raster/ Vector/ Hybrid**
7. New 2020 Camera Dialogue allows you to select from Standard Views (Top/ Sides/ Bottom.
8. Orthographic/ Perspective toggle on/ off.
9. Scale selector (This automatically turns on "Preserve Scale on Resize" when used. This can also be turned on/ off.)
10. New 2020 Effects Dialogue: Fog: Colour/ On/ Off. Includes Date & Time*
11. New 2020 Styles Dialogue is a separate section instead of being a tab. Here you can choose from the set of Styles, whether they're from "In Model Styles" group or from Standard Collections- Just click on the flyout to see what collections are available You can also create your own Styles Collections from within SketchUp.
12. New 2020 Tags Dialogue: Now you can control what Tags (Formerly called "Layers") that you can see or hide within the Model Window. This means you no longer have to completely set up everything in your SU model.

CONSTRUCTION DOCUMENTS USING SKETCHUP PRO 2020

DRAWING PRODUCTION: LAYOUT
SCALED 2D DRAWING

Clicking on the **Scaled Drawing Dialogue** activates the "Make a Scaled Drawing" button. Clicking on this button opens a window that allows you to draw according to your chosen scale.

Draw to scale in LayOut:

1. Click on "Make Scaled Drawing" Button.
2. Choose Scale (For example 1 : 10), Length Type and Units. The various options are illustrated below.
3. Click on the Line tool.
4. Click on a start point and point your cursor in the direction you want to draw.
5. Type a number and hit "Enter". The resulting line will be appropriately dimensioned.
6. Dimensions will automatically reflect scale.
7. The Join Tool binds together lines or curves whose endpoints meet.
8. The Split Tool breaks apart lines, curves or shapes into separate segments.
9. The Select tool enables positioning of lines but also enables editing of shapes by double-clicking.
10. Dimensions applied will reflect your scale automatically.
11. Lines and infills are edited as previously illustrated.
12. To finish, click Escape.
13. The drawing is a discrete Group. To edit, use Select Tool and double click.

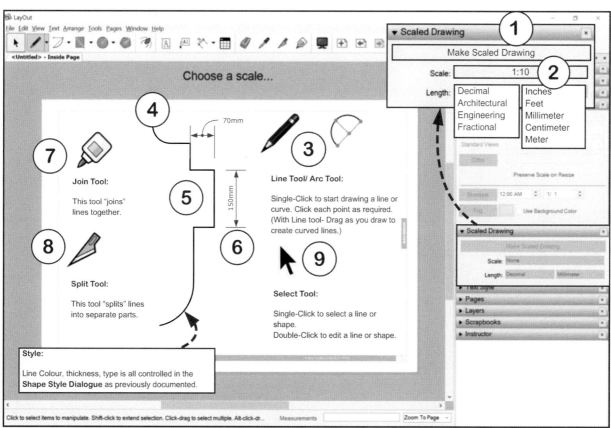

CONSTRUCTION DOCUMENTS USING SKETCHUP PRO 2020

DRAWING PRODUCTION: LAYOUT
PAGES & LAYERS

The Pages Dialogue displays the order of pages and provides an interface for naming them. The order of pages can be controlled by dragging and dropping. Page names and numbers can be used for "Auto-Tagging" so that one tag can be used across multiple pages avoiding the need to individually number and name each sheet.

The Dialogue contains:

1. The Sequence (Order) of Pages
2. Page Numbers
3. Page Names
4. Include in Presentation On/ Off
5. Add Page Button
6. Duplicate Page Button
7. Delete Page Button
8. Display as List Button
9. Display as Grid Button

The Hash denotes the page at which the page numbers start. To access page numbers follow this menu: File > Document Setup > Autotext > Page Number.

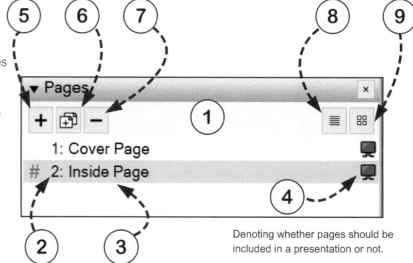

Denoting whether pages should be included in a presentation or not.

The Layers Dialogue displays the order in which different elements are displayed. Those higher up in the sequence are "in front of" those lower down. The order can be changed by drag-drop.

Features of the Layers Dialogue:

10. Current (Active) layer is denoted by the Pencil icon and the blue background.
11. Layer of any selected object is denoted by a blue square.
12. "Eye" icon denotes whether a layer is visible on a particular page.
13. Lock Icon "locks" a layer so that it can't be edited.
14. The "shared layer" icon signifies layers that are visible across multiple pages.

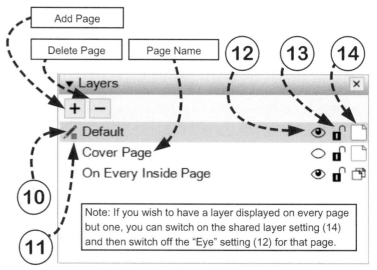

CONSTRUCTION DOCUMENTS USING SKETCHUP PRO 2020

DRAWING PRODUCTION: LAYOUT

SETTING OUT YOUR SHEETS

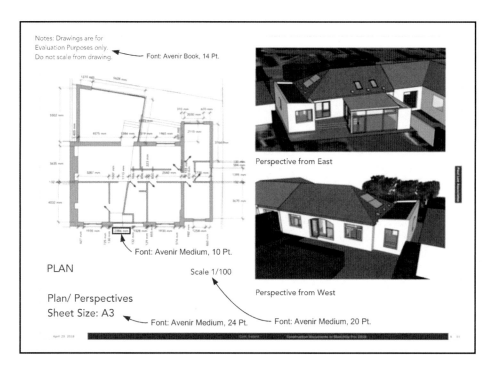

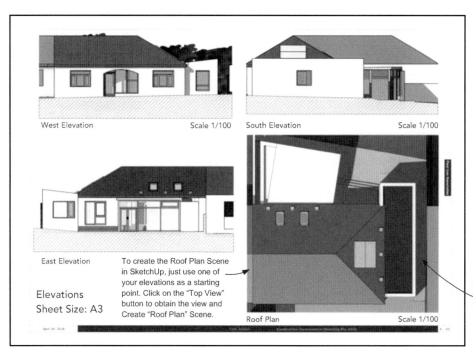

Finishing the Sheets

As we saw previously, drawings are inserted via:

File > Insert > (SketchUp File)

The SketchUp window is inserted.

To set up our Plan:
Right-click on the SketchUp Window and select the Scene (Here it's the "Plan_0" Scene that we created.)

Right-click again on the SketchUp Window and select the scale (1:100 in this case.)

Drag the sides and corners of the window into position so you can see the whole drawing.

Drag on the model to position the SketchUp Window.

Holding down the CTRL key (or alt for Mac), drag the SketchUp Window to create a copy.

If you need to edit the Scene, go back to your SketchUp model by right-clicking on it and selecting **"Open with SketchUp"**.

You can then make changes to the model or create a new Style for that Scene. Remember:

1. In SU: Update the Style
2. In SU: Update the Scene
3. In SU: Save the SketchUp File
4. In LO: Right-click on the SketchUp Window and select "Update Model Reference"

"Top View" button in SketchUp.

CONSTRUCTION DOCUMENTS USING SKETCHUP PRO 2020

DRAWING PRODUCTION: LAYOUT
SCRAPBOOK

Scrapbook is a kind of LayOut "browser" which allows you to "click and drop" graphics from the window into your LayOut Page. Scrapbook contains lines, text, shapes, drawings of trees, cars, people arrows and many other things that you can use to instantly populate your drawings with information.

1. Drop-down menu with sub-menus.
2. Graphics for selection.
3. Next Scrapbook Page/ Previous Scrapbook Page buttons.
4. Edit current "Page" (Opens this as a LayOut file.)
5. Click-Select and Click-Place objects.

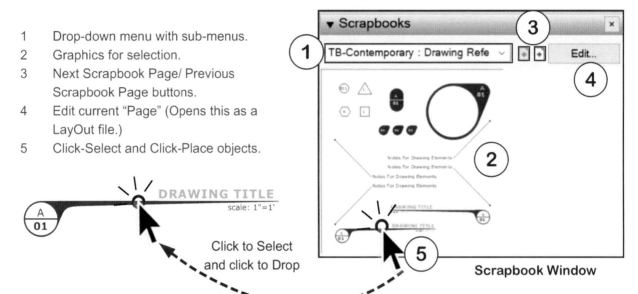

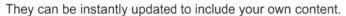

Some of the selection of graphics available for insertion.

Scrapbooks contain both SketchUp files and simple line graphics.
They can be instantly updated to include your own content.
Select cars, trees and other graphics of various scales.

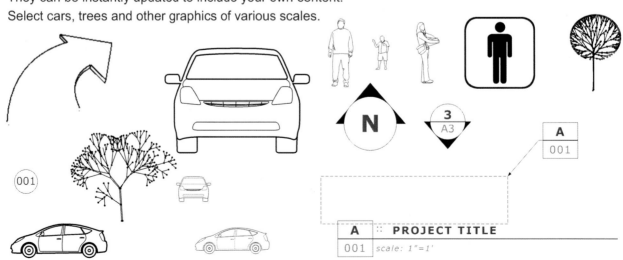

Select cars, trees and other graphics of various scales.

CONSTRUCTION DOCUMENTS USING SKETCHUP PRO 2020

DRAWING PRODUCTION: LAYOUT

THE SECTION SHEET

Previously we saw how to import, position and size our SketchUp files. (Refer to "Drawing Production" Page 103) Below we have the typical sizes and settings for an A3 size drawing sheet at 1:50 size.

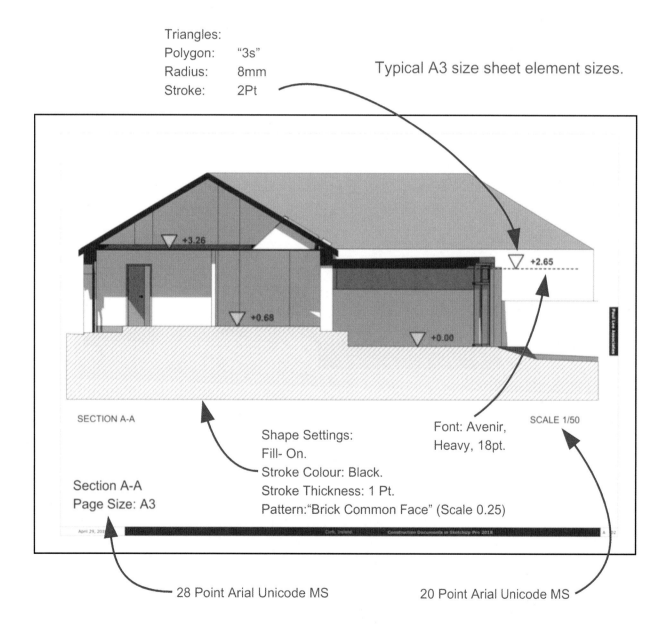

CONSTRUCTION DOCUMENTS USING SKETCHUP PRO 2020

LAYOUT DETAILS

SET UP THE DETAILS LAYOUT SHEET

Below is the Details Sheet showing the recommended sizes of Text Fonts, Scenes and graphics. Let's Insert our SketchUp File Scenes: "Perspective" and "Section".

Go to: File > Insert [The Details Model File].
The SketchUp window should appear.
Right-click on the SketchUp window and select the "Perspective" Scene.
Using Shift-Drag, Copy the SketchUp window then right-click to select the "Section" Scene.
Position and size your window as follows   (We won't set the Scale just yet)

In the **SketchUp Model** Window, the "**Preserve Scale on Resize**" tickbox allows you to resize the model using the model Window size, or to freeze the scale.

1. Drag the window into position at top-left.
2. Drag the bottom-right corner across to fill the intended area.
3. Once you have the size you want, you can freeze the Scale in the SketchUp Model Dialogue to implement the Window size.
4. To adjust the scale and position of the Perspective: Double-click into the window and use your mouse to zoom and pan (Use the same controls that are used for SketchUp.)

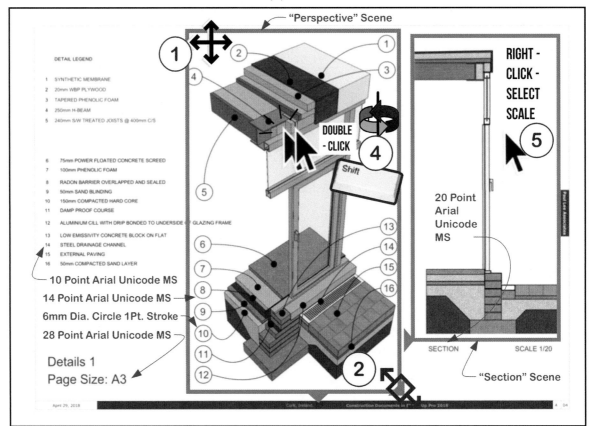

MODELLING THE HOUSE: FINISHUP

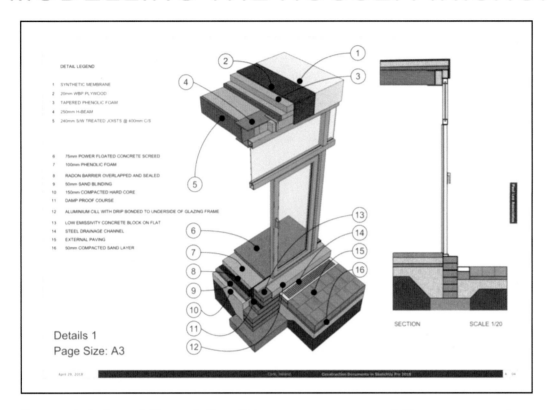

Congratulations! Completing this course successfully means you're now able to produce professional documents in SketchUp Pro.

If you have any questions, please don't hesitate to get in touch- **paul@sketchup.expert** or visit sketchup.expert for more information about training and advice.

Sign up for exclusive offers for future training products and material & model downloads. Visit **sketchup.expert** and click on the **Resources Page**.

Thanks for buying my book!

I'll be adding features to the website with freebies and downloads. Make sure to sign up with the word "SketchUp Constructor" in the message or email header for exclusive access.
Also, join the discussion thread about SketchUp for Construction & Working Drawings Discussion in SketchUcation here:
https://sketchucation.com/forums/viewtopic.php?f=12&t=15911

CONSTRUCTION DOCUMENTS USING SKETCHUP PRO 2020

ACKNOWLEDGMENTS

I'd like to thank Mike Lucey and Rich O'Brien of SketchUcation for their encouragement in producing SketchUp material throughout the years. Also thanks to the folks at SketchUp for their interest, most especially Josh Reilly and John Bacus. Aaron Dietzen's videos are always enlightening and great for keeping up to date with the latest tech improvements. Thanks also to Allyson McDuffie for helping out with some of our educational projects in Ireland.

Aidan Chopra was extremely helpful to me during his SketchUp days, before he went to co-found Bitsbox. Omar-Pierre Soubra of Trimble was really helpful with a number of laser scan-to-SketchUp projects .

I'd also like to acknowledge some of the great SketchUp trainers and writers such as Bonnie Roskes, Mike Brightman, Matt Donley, Mike Tadros, Alex Oliver of SketchUpSchool, Nick Sonder, Dave Richards and Daniel Tal.

Thanks to Nick Johnson of Cadsoft Solutions Ltd. UK for his kind assistance.

SketchUp has changed the nature of my profession for the better. Those who have switched over from 2D CAD I'm sure will agree. I'd like to thank the inventors. I haven't met Brad Schell but would very much like to shake his hand. Thank you to Brad and all those who created this marvellous software.

I'd like to acknowledge IvanOnTech.com Academy* for excellent cryptocurrency educational resources and YouTube channel which have already helped me with my commercial and financial skills.

I'd like to thank my two wonderful kids Rob and Anna for being such great supporters (and distractors just when needed). Also parents Teresa and Mahon and brothers John and Mark.

*Affiliate Link Main Academy Offer with Full Course Access (Trial + Paid Membership) https://academy.ivanontech.com/a/17936/9FMYTjmH

Made in the USA
Coppell, TX
21 September 2020